Handbook of
IoT and Big Data

Science, Technology, and Management Series

Series Editor
J. Paulo Davim

This book series focuses on special volumes from conferences, workshops, and symposiums, as well as volumes on topics of current interest in all aspects of science, technology, and management. The series will discuss topics such as mathematics, chemistry, physics, materials science, nanosciences, sustainability science, computational sciences, mechanical engineering, industrial engineering, manufacturing engineering, mechatronics engineering, electrical engineering, systems engineering, biomedical engineering, management sciences, economical science, human resource management, social sciences, and engineering education. The books will present principles, models techniques, methodologies, and applications of science, technology, and management.

Advanced Mathematical Techniques in Engineering Sciences
Mangey Ram and J. Paulo Davim

Optimizing Engineering Problems through Heuristic Techniques
Kaushik Kumar, Nadeem Faisal, and J. Paulo Davim

Handbook of
IoT and Big Data

Edited by
Vijender Kumar Solanki
Vicente García Díaz
J. Paulo Davim

CRC Press
Taylor & Francis Group
Boca Raton London New York

CRC Press is an imprint of the
Taylor & Francis Group, an **informa** business

CRC Press
Taylor & Francis Group
6000 Broken Sound Parkway NW, Suite 300
Boca Raton, FL 33487-2742

Library of Congress Cataloging-in-Publication Data

Names: Solanki, Vijender Kumar 1980- editor. | García Díaz, Vicente, 1981- editor. | Davim, J. Paulo, editor.
Title: Handbook of IoT and big data / edited by Vijender Kumar Solanki, Vicente García Díaz, and J. Paulo Davim.
Description: Boca Raton : Taylor & Francis, 2019. | Series: Science, technology, and management series | Includes bibliographic references.
Identifiers: LCCN 2018047209| ISBN 9781138584204 (hardback : alk. paper) | ISBN 9780429053290 (ebook)
Subjects: LCSH: Internet of things--Handbooks, manuals, etc. | Big data--Handbooks, manuals, etc. | Embedded Internet devices--Handbooks, manuals, etc.
Classification: LCC TK5105.8857 .H34 2019 | DDC 004.67/8--dc23
LC record available at https://lccn.loc.gov/2018047209

Visit the Taylor & Francis Web site at
http://www.taylorandfrancis.com

and the CRC Press Web site at
http://www.crcpress.com

Contents

Preface

The Internet of Things (IoT) is emerging as one of the most prominent technological concepts in the twenty-first century. We live in a world where new objects are connected to the Internet with the aim of improving and facilitating people's lives. In addition, objects that we did not even imagine a few years ago as part of the network of networks are now beginning to be connected, offering a new immense range of possibilities.

All this is generating a quantity of data that grows exponentially and that, thanks to new technologies and new software platforms, can be processed with speed that was unattainable only a few years ago. However, the billions of existing objects and the enormous amounts of data that derive from them make necessary an active collaboration of the scientific community in these so heterogeneous fields, with the objective of improving different aspects.

Some of the current lines of work include standards and communication protocols; regulations; simulations; or retrieval, storage, processing, and visualization of the data itself. That is why this book *Handbook of IoT and Big Data* has been written with the collaboration of international scientists who provide different visions and solutions to some of the most important current problems from the point of view of the Internet of Things community, from the point of view of the Big Data community, and from an even more interesting mixed point of view.

Chapter 1 is entitled "Big Data Analysis and Compression for Indoor Air Quality." In this chapter, the authors focus on the gradually growing concern about safe environments. Indoor air quality is an important concern because it leads to big health and economic problems. The study collects a great amount of data from different sources and then applies the ubiquitous Snappy data compression technique to be compared to other available techniques to compress data obtained in different contexts.

Chapter 2 is entitled "Programming Paradigms for IoT Applications: An Exploratory Study." The authors explain the importance of the Internet of Things in smart cities environments. However, they state that there is a gap between users and IoT applications from a programming perspective. To address the problem, adaptive programming together with cloud technologies are proposed by using six exploratory case studies through simulations using different technological platforms like ThingSpeak.

Chapter 3 is entitled "Design and Construction of a Light-Detecting and Obstacle-Sensing Robot for IoT—Preliminary Feasibility Study." This chapter introduces the design and construction of a light-sensing and obstacle-detecting robotic vehicle using a microcontroller. It provides an infrastructure with obstacle-sensing capabilities together with an IoT-based architecture by using features such as photo detector, ultrasonic sensor, and light.

Chapter 4 is entitled "XBee and Internet of Robotic Things Based Worker Safety in Construction Sites." The focus of this chapter is on construction sites, which, according to the authors, are considered to be very risky and dangerous places. The chapter proposes a high-tech model for a system with the aim of gathering information and giving alarms in dangerous zones. To accomplish that task, the authors rely on the IoT to monitor and collect a range of data using different sensors with the help of XBee devices.

Chapter 5 is entitled "Contribution of IoT and Big Data in Modern Health Care Applications in Smart City." In this chapter, the authors talk about the emerging Web of Things, in which the linked computing devices are associated with the Web as a new type of resource. In addition, the authors focus on the problem of security and privacy of the devices integrated with the Web, especially highlighting the application-based security issues. To address the problem, solutions and related technologies are reviewed and presented.

Chapter 6 is entitled "Programming Language and Big Data Applications." In this chapter, the authors address the problem of processing the large amount of structured, semi-structured, and unstructured data emerging with new technologies. The chapter focuses on providing a review of some challenges, emerging trends, and programming languages, and a global view of big data technologies that can be selected according to technological needs and specific applications' requirements.

Chapter 7 is entitled "Programming Paradigm and the Internet of Things." According to the authors, there are several concerns regarding the programming paradigm on IoT, like heterogeneity caused by hardware, software and communication platforms, volume of generated data, various forms of generated data, and more. To mitigate this, the authors focus and explain the "polyglot programming" approach, consisting of the integration platform and platform-supported languages.

Chapter 8 is entitled "Basics of the Internet of Things (IoT) and Its Future." This is a summary chapter in which the authors focus on some of the most interesting concepts behind the IoT like the different possible applications of the IoT, the components related to IoT architectures, the importance and type of sensors, the main network technologies, the protocols and standards, the most relevant current challenges, and some possible lines of future work.

Chapter 9 is entitled "Learner to Advanced: Big Data Journey." In this chapter, the authors talk about a few machine learning algorithms together with some big data sources, from the point of view of a business context: recognizing methods, approaches, and tools used to transform large amounts of available data into valuable information, and supporting business decision making by applying machine learning algorithms. It also focuses on the latest technologies used in the Hadoop platform.

Chapter 10 is entitled "Impact of Big Data in Social Networking through Various Domains." In this chapter, the authors present a cloud-based architecture that relies on the MapReduce concept. It also includes analytics mechanisms that can be used to detect and solve malicious activities in public social networks. The authors also analyze the impact of data growth in various domains in which a range of different sensors are commonly used in the IoT.

Chapter 11 is entitled "IoT Recommender System: A Recommender System Based on Sensors from the Internet of Things for Points of Interest." In this chapter, the authors introduce a novel recommendation system to generate recommendations for users based on both the information obtained from common sensors and the opinions explicitly indicated by users about places or recommendations previously made by the system.

Chapter 12 is entitled "Internet of Things: A Progressive Case Study." In this chapter, the authors present a smart light system that offers an easy, reliable, and efficient way of handling and controlling lights through a wireless medium and remote access. The system replaces the conventional wired method, which is costly and requires management of a lot of cable connections. Thus, the fault detection, maintenance, and management of smart light systems are comparatively easy and efficient.

Chapter 13 is entitled "Big Data and Machine Learning Progression with Industry Adoption." This chapter presents a study that is intended to explain how machine learning is changing business dynamics using big data, how companies can embrace the new technologies, what are the inputs that could be required, and so on. The study also discusses the foundation of machine learning followed by the importance of machine learning algorithms through well-known business applications.

Chapter 14 is entitled "Internet of Things: Inception and Its Expanding Horizon." In this chapter, the authors deal with different technical and nontechnical definitions of the Internet of Things and other interesting aspects such as a historical view of its development. They also deal with some common applications of the related technologies together with the benefits and the expected growth of the IoT. To conclude, the authors introduce a possible future scenario.

Chapter 15 is entitled "Impact of IoT to Accomplish a Vision of Digital Transformation of Cities." In this chapter, the authors present a review of IoT issues and challenges in the emerging stages of the design of smart cities in India. Some of the issues addressed by the authors include legal, regulatory, economic, infrastructure, security, and privacy aspects. The study also deals with network communication models used in establishing connection between devices and the Internet.

The editors are thankfully indebted to Almighty God for giving us the blessing to complete this book. Completing a book is not an easy task; it starts with consuming many hours, months, and even a year or more. We can attest that during that time, we were working closely with publishers, editors, and authors. We are very much thankful to our beloved Professor J. Paulo Davim, series editor, CRC Press, the Taylor & Francis Group, for having faith and giving us a chance to edit this book. From proposal submission till completion, your support is very helpful to us. We are sincerely thankful to the CRC Press, Taylor & Francis Group members, Cindy Renee Carelli, and Ms. Erin Harris for your timely support in liaising with us. We wanted to convey our thanks to all authors who participated in this book project; due to scope and quality, we could not accept many good chapters, but we are sure that the work included in this book will prove helpful to the young researchers and industry entrants in the field of IoT and big data. Our book will help them to solve their planning strategy and will proved the boost to solve their real-time problem. In the end we are very much thankful to our institution principals—CMR Institute of Technology (Autonomous); Hyderabad; TS India; University of Oviedo, Spain; and University of Aveiro, Portugal—for giving us an open-research platform to complete this book in time.

We will be very happy to hear your feedback about this book. Though utmost care has been taken to select the chapters and authors' work, which was closely monitored and revised with rigorous peer review, authors' or readers' reviews and feedback will be very useful for us, to ensure their points are addressed in our forthcoming books.

We request you to purchase this book for your institution library and research lab, and to take advantage of cutting-edge technology exploration information throughout this book.

Vijender Kumar Solanki, PhD
CMR Institute of Technology (Autonomous)
Telangana, Hyderabad, India

Vicente García Díaz, PhD
University of Oviedo, Spain

J. Paulo Davim, PhD, D.Sc
University of Aveiro, Portugal

Editors

Vijender Kumar Solanki, PhD, is an Associate Professor in Computer Science & Engineering, CMR Institute of Technology (Autonomous), Hyderabad, TS, India. He has more than 12 years of academic experience in network security, IoT, big data, smart cities and IT. Prior to his current role, he was associated with the Apeejay Institute of Technology, Greater Noida, UP; KSRCE (Autonomous) Institution, Tamilnadu, India; and Institute of Technology and Science, Ghaziabad, UP, India.

He has attended an orientation program at UGC–Academic Staff College, University of Kerala, Thiruvananthapuram, Kerala, and a refresher course at the Indian Institute of Information Technology, Allahabad, UP, India.

He has authored or coauthored more than 26 research articles that are published in journals, books, and conference proceedings. He has edited or coedited four books in the area of Information Technology. He received a PhD in Computer Science and Engineering from Anna University, Chennai, India, in 2017; an ME, MCA from Maharishi Dayanand University, Rohtak, Haryana, India, in 2007 and 2004, respectively; and a bachelor's degree in Science from JLN Government College, Faridabad Haryana, India, in 2001.

He is the Editor of the *International Journal of Machine Learning and Networked Collaborative Engineering* (IJMLNCE) ISSN 2581-3242, Associate Editor of the *International Journal of Information Retrieval Research* (IJIRR), IGI-GLOBAL, USA, ISSN: 2155-6377 | E-ISSN: 2155-6385. He is a guest editor with IGI-Global, USA, InderScience, and many more publishers. He can be contacted at spesinfo@yahoo.com.

Vicente García Díaz is a Software Engineer and has a PhD in Computer Science. He is an Associate Professor in the Department of Computer Science at the University of Oviedo. He is also part of the editorial and advisory board of several journals, and has been the editor of several special issues in books and journals. He has supervised more than 70 academic projects and published more than 70 research papers in journals, conferences, and books. His research interests include machine learning, domain-specific languages, and electronic learning.

J. Paulo Davim received his PhD degree in Mechanical Engineering in 1997, M.Sc. degree in Mechanical Engineering (materials and manufacturing processes) in 1991, and Mechanical Engineering degree (5 years) in 1986, from the University of Porto (FEUP) the Aggregate title (Full Habilitation) from the University of Coimbra in 2005; and the DSc from London Metropolitan University in 2013. He is EUR ING by FEANI–Brussels and Senior Chartered Engineer by the Portuguese Institution of Engineers with a MBA and

Specialist title in Engineering and Industrial Management. Currently, he is Professor at the Department of Mechanical Engineering of the University of Aveiro, Portugal. He has more than 30 years of teaching and research experience in Manufacturing, Materials, Mechanical and Industrial Engineering with special emphasis in Machining and Tribology. He also has interest in Management, Engineering Education, and Higher Education for Sustainability. He has guided large numbers of postdoc, PhD, and masters students as well as coordinated and participated in several financed research projects. He has received several scientific awards. He has worked as evaluator of projects for international research agencies as well as examiner of PhD theses for many universities. He is the Editor in Chief of several international journals, Guest Editor of journals, books Editor, book Series Editor, and Scientific Advisory for many international journals and conferences. Presently, he is an Editorial Board member of 25 international journals and acts as reviewer for more than 80 prestigious Web of Science core collection journals. In addition, he has also published as editor (and coeditor) more than 100 books and as author (and coauthor) more than 10 books, 80 book chapters, and 400 articles in journals and conferences (more than 200 articles in journals indexed in Web of Science core collection/h-index 45+/6500+ citations and SCOPUS/h-index 53+/8500+ citations).

Contributors

Shivani Agarwal
KIET School of Management
KIET Group of Institutions
Ghaziabad, India

Rohit Anand
Electronics & Communication
  Engineering
G.B. Pant Engineering College
New Delhi, India

Nabila Ansari
Faculty of Management Science
Gurukul Kangri Vishwavidyalaya
Haridwar, Uttarakhand, India

Sivadi Balakrishna
Department of CSE
Pondicherry Engineering College
Pondicherry University
Puducherry, India

Sourav Banerjee
Department of Computer Science
  and Engineering
Kalyani Government Engineering
  College
West Bengal, India

Swagat Rameshchandra Barot
Department of Mathematics
R. C. Technical Institute
Ahmedabad, Gujarat, India

Nikita Bhatt
U & P U. Patel Department of
  Computer Engineering
CSPIT, Charusat
India

Chinmay Chakraborty
Electronics & Communication
  Engineering
Birla Institute of Technology
Jharkhand, India

J. Paulo Davim
University of Aveiro
Aveiro, Portugal

Mohammad Farhan Ferdous
School of Information Science
Japan Advanced Institute of Science
  and Technology
Nomi, Japan

Amit Ganatra
U & P U. Patel Department of
  Computer Engineering
CSPIT
Charusat, India

Cristian González García
Model-Driven Engineering
  Research Group
Department of Computer Science
University of Oviedo
Oviedo, Spain

Vicente García Díaz
Model-Driven Engineering
  Research Group
Department of Computer Science
University of Oviedo
Oviedo, Spain

Anita Gehlot
School of Electronics & Electrical
   Engineering
Lovely Professional University
Jalandhar, India

Subhanshu Goyal
Department of Mathematics
Marwadi University
Rajkot, Gujarat, India

Divyanshu Gupta
School of Electronics & Electrical
   Engineering
University of Petroleum and Energy
   Studies
Dehradun, India

Meenu Gupta
School of Computer Science
Lingaya's Vidyapeeth Faridabad
India

Mihir Joshi
Faculty of Management Science
Gurukul Kangri Vishwavidyalaya
Haridwar, Uttarakhand, India

Abhishek Kumar
Department of Computer Science
   and Engineering
Aryabhatta College of Engineering
   & Research Center
Ajmer, India

Anil Kumar
Department of Computer Science
Deen Dayal Upadhyaya College
New Delhi, India

Somesh Kumar
Department of Information
   Technology
Noida Institute of Engineering and
   Technology
Greater Noida, Uttar Pradesh, India

Sandhya Makkar
Department of Operation
   Management
Lal Bahadur Shastri Institute of
   Management
Dwarka, New Delhi, India

Aditee Mattoo
Department of Information
   Technology
Noida Institute of Engineering and
   Technology
Greater Noida, Uttar Pradesh, India

Daniel Meana-Llorián
Model-Driven Engineering
   Research Group
Department of Computer Science
University of Oviedo
Oviedo, Spain

Edward Rolando Núñez-Valdez
Model-Driven Engineering
   Research Group
Department of Computer Science
University of Oviedo
Oviedo, Spain

Srinivas Kumar Palvadi
Department of Computer Science
   and Engineering
Sri Satya Sai University of
   Technology & Medical Sciences
Sehore, Madhya Pradesh, India

Sudipta Paul
Department of Computer Science
Kalyani Mahavidyalaya, India

V. B. Surya Prasath
Department of Electrical
   Engineering and Computer
   Science
University of Cincinnati
Cincinnati, Ohio

Geeta Rana
Himalayan School of Management
  Studies
Swami Rama Himalayan University
Dehradun, India

Mamata Rath
Department of Computer Science &
  Engineering
C.V. Raman College of Engineering
Bhubaneswar, India

Pramod Singh Rathore
Aryabhatta College of Engineering
  & Research Centre
Ajmer, India

T. V. M. Sairam
Department of Computing Science
  and Engineering
Vellore Institute of Technology
Chennai, Tamil Nadu, India

Khushboo Sansanwal
Computer Science & Engineering
G.B. Pant Engineering College
New Delhi, India

Kavita Sharma
Department of Computer
  Engineering
National Institute of Technology
Kurukshetra, India

Neeraj Sharma
Amity University
Kolkata, India

Ravindra Sharma
Himalayan School of Management
  Studies
Swami Rama Himalayan University
Dehradun, India

Gulshan Shrivastava
Computer Science and Engineering
National Institute of Technology
Patna, India

Neha Singla
Department of Computer Science
Punjabi University
Patiala, India

Rajesh Singh
School of Electronics & Electrical
  Engineering
Lovely Professional University
Jalandhar, India

V. K. Singh
Faculty of Management Science
Gurukul Kangri Vishwavidyalaya
Haridwar, Uttarakhand, India

Vijender Kumar Solanki
CMR Institute of Technology
  (Autonomous)
Hyderabad, India

M. Thirumaran
Department of CSE
Pondicherry Engineering College
Pondicherry University
Puducherry, India

Prayag Tiwari
Department of Information
  Engineering
University of Padova
Padua, Italy

Shagun Tyagi
Faculty of Management Science
Gurukul Kangri Vishwavidyalaya
Haridwar, Uttarakhand, India

1

Big Data Analysis and Compression for Indoor Air Quality

Khushboo Sansanwal, Gulshan Shrivastava,
Rohit Anand, and Kavita Sharma

CONTENTS

1.1 Introduction

There have been several works around the world to define what constitutes the Quality of the Indoor Air (IAQ). In one such work, "Ventilation for Acceptable Indoor Air Quality," Quality of the Indoor Air has been defined as "air in which there are no known contaminants at harmful concentrations as determined by cognizant authorities and with which a substantial majority [80% or more] of the people exposed do not express dissatisfaction" (Persily 1997).

Further, the Quality of the Indoor Air has been defined by the World Health Organization (WHO) as "the physical and chemical nature of indoor air, as delivered to the breathing zone of building occupants, which produces a complete state of mental, physical and social well-being of the occupants, and not merely the absence of disease or infirmity" (Heseltine and Rosen 2009).

Over the last few decades, the importance of the quality of indoor air inside buildings, office spaces, schools, and so on has drawn a significant concern, as there has been an undeniable accumulation of evidence that the deterioration or contamination of the air quality inside the buildings is leading to health problems among the occupants (Heinsohn and Cimbala 2003). The researchers have further proved that mitigation of the air quality has brought substantial improvement in the health symptoms, well-being, and comfort of the occupants of the buildings, in turn leading to increased productivity (Alker et al. 2014). This has thus even justified that improving the same makes economic sense by increasing the productivity on one side and decreasing the number of finances spent on health restoration of the occupants arising on account of poor indoor air quality. It has been seen that there is a compelling reason for the managers of affected buildings to address this issue, as air quality deterioration and contamination are occupational hazards also (Kephalopoulos et al. 2007, Solanki et al. 2015, Kadam et al. 2019, Dhall and Solanki 2017, Solanki et al. 2018a, 2018b).

As found by the United States Environmental Protection Agency, substantial evidence exists for showing urgent concern for the care of quality in the indoor air, as serious risks have been found to be there for the inhabitants of such affected buildings whenever quality of the air indoors has deteriorated. Further such deterioration of indoor air has been found to be existing as an essential risk factor among the top five risks for the people's health (Jones 1999). Thus, immediate and appropriate action is required to ensure the quality maintenance of the indoor air to avoid any catastrophic consequences on the people's health (Roulet et al. 1994). In 2009, the World Health Organization (WHO) presented their work on the risks faced by the health of the people worldwide. The said extensive work established firmly that deterioration in the quality of the indoor air leads to diseases and hence was found to account for around 3% of the diseases faced by the world, a number which is too vast to be ignored (WHO 2017).

Further research has shown that there has been ample evidence to suggest that the overall productivity of the inhabitants of the buildings possessing contaminated indoor air gets jeopardized. Several effects were found to be taking a toll on such affected inhabitants, causing absence from work, leaves taken for the impact on health, decrease in productivity and efficiency in the work done, and so on (Alker et al. 2014). The Occupational Safety and Health Administration, an organization of the United States, in an another work found the impact on the profitability of business to have suffered at least 15 minutes of loss overall in the business generated by a worker over each day of work, on account of the contaminated air indoors (Occupational Safety and Health Administration 2015). The loss is further compounded by the medical costs of the restoration of the health of such affected employees (Nathanson 1993). When estimates have been tried to be made on the loss suffered in the performance of the work done in another study, a substantial loss on account of performance, to the extent of around 4%, has been found due to the contaminated air indoors (US EPA, I-BEAM 2008).

It is understandable that a considerable investment is required to be made to avoid the contamination/deterioration of the indoor air and thus is a matter of concern for the finances in managing such buildings but for a relief it has been found that the expenditure made on such account, in turn, returns the cost incurred in the form of overall improvement, preservation of the health of workers, further leading to preserving the performance of workers (Occupational Safety and Health Administration 2015).

In yet another study being done by the EPA, the loss on productivity was estimated, and it was found that productivity can suffer a decrease of around 5% on account of the deterioration of the quality of air indoors (NEMI 1994). It was again shown that the cost incurred on account of actions made about the indoor air was far less than the cost incurred for the preservation of the health of the inhabitants of such buildings. Further, it has been found that the cost incurred on the buildings was far less, in fact, less than one fourth to that which was spent on a worker of such buildings every year. It clearly showed the loss suffered on account of productivity: a loss as meager as 1% leads to a cost far more than what would have been suffered on account of maintenance of such buildings (Spengler 2001). About the buildings and their internal environments' effect on the health of the occupants of the same, there are two types of illness-related effects. The first is building-related disease, and the second is a nonspecific building-related illness that is universally acknowledged with a nomenclature of "Sick Building Syndrome." This disease has a syndromal presentation whereby the occupants of the building are presented with the development of symptoms such as irritation in the eyes, stuffiness of the nose, respiratory symptoms and illnesses, running nose, headache, and alteration in awareness (Jansz 2011). The symptoms occur in the affected occupants either due to the introduction of noxious chemicals or due to an accumulation of noxious chemicals on account of poor ventilation. Also, these symptoms may occur due to the growth of the disease-producing organisms in the dampness on the walls, such as the growth of mold or fungus. As may be seen in the preceding discussion, these offending substances may be either nonbiological (e.g., chemical) or biological. These illnesses sometimes may be transient, recovering in due time with the rectification of the individual causes. In worse cases, due to long-term exposure to molds and fungus, severe respiratory diseases may occur such as allergic pneumonitis (Sahlberg 2012). At times, such a disease may cause irreversible damage, in which the withdrawal of the causative agent may no longer lead to the recovery of the body. This leads to the importance of the rectifications in the quality of the air indoors and also the importance related to the implementation of rectifying measures in time to avoid such irreversible damage. Building-related disease, by comparison, is a dangerous and severe condition. In building-related illnesses, there are established clear diagnostic criteria in which recognized and research-proven path-physiological characteristics and parameters are used to identify the health problems. Such health illnesses occur due to development of the lesions due to the exposure to the biological; that is. infective, immunologic agents or chemicals such as toxins and irritants. The examples of some of the diseases are asthma and pneumonitis due to hypersensitivities, and inflammation of the airways such as laryngitis, bronchitis, sarcoidosis, and dermatitis. Further, there may be poisonings due to gases such as carbon monoxide and radon (Apter et al. 1994).

Maintenance and preservation of excellent indoor air quality can effectively contribute to the positive health outcomes of the occupants of such buildings. The positive change is brought by the favorable health promoting environment thereby increasing the comfort of the occupants and in turn, increasing the overall productivity of the people (Nathanson 1993).

The increase in the people's productivity is a significant increase which has been well supported by the available research work. Further, the research data has substantially shown that the buildings where the air quality has been better are concerned with the fewer workers being seen with evidence of the syndromal presentation of the disease called "sick building syndrome" and thus savings on the expenditure of health makes the investment in the better indoor air quality measures a prudent one (Apter et al. 1994; Occupational Safety and Health Administration 2015).

This research study aims to investigate the understanding of the big data related to the quality of indoor air of buildings as well as its management through compression techniques. There are numerous studies on indoor air quality, big data analysis, and big data compression. Still, only a very few researchers combined the concept of indoor air quality analysis and its compression. For example, Maitrey and Jha (2015) investigated a cloud technology, MapReduce, for the big data analysis. Cociorva and Iftene (2017) presented the evaluation of indoor air quality for the smart control of the different kinds of intelligent devices. Yang et al. (2014) suggested an individual compression and scheduling based approach for the excellent processing of big data in the cloud. In this research study, the indoor air quality big data analysis has been done with a suitable compression technique called Snappy. The second contribution of this chapter is that it helps to validate the Snappy compression technique against the other well-known compression techniques with the help of the parameters like time taken and compression ratio. In addition, it also aims to provide useful insights helpful for managing, storing, processing, and analyzing the big data.

1.2 Sources of Indoor Air Pollutants

Various contaminants and pollutants can lead to a fall in the quality of the indoor air. The hazardous nature of individual contaminants and the extent of the power of the ventilation modules determine the extent to which there may be an impact on the quality of the air indoors. Also, the maintenance done on the individual machines and the age of such machines have a determining potential on the extent of deterioration in the air quality caused. Other factors may be the local weather, the plan of the building, site, location of the building, the age of the building, what maintenance is done, use of the materials, and so on. For quite some time, the potential sources of air pollutants in the buildings have increased manyfold (Sahlberg 2012). Some other pollutants are as follows (Lee et al. 2002):

1. The electronics used in the offices such as computer machines and associated electronic accessories.
2. The printers and photocopiers.
3. The hardware used for photography and developing prints.
4. Various chemical substances used in the buildings: for instance, glues, thinners, paints, and boards have the potential to emanate formaldehyde.
5. The use of several types of plastics and polyvinyl chlorides.
6. The detergents and cleaning materials containing several types of offending chemicals.
7. The exhaust emanated from vehicles, such as in indoor parking situated in basements of buildings.
8. The exhaust expelled from chimneys.

1.3 Categories of Indoor Air Pollutants

Among the various pollutants that can affect the quality of the indoor air, classification for the several parameters of the pollutants can be made. In one such classification, the pollutants have been classified depending upon (a) the origin from life, that is, biological; (b) the constitution, that is, chemical; and (c) the size of the particles (Persily 1997; Lister et al. 1998).

Among the categories of biological pollutants, the hazardous presence of bacteria, fungus, and viruses are the most important ones. Poor dust control may lead to the presence of mites with the attending effects on the health of the inhabitants. Inadequate controls over humidity inside the building, the dampness on account of seepage, poor sanitary conditions, and spillage of sewage through corroded pipes contribute heavily to the presence of biological pollutants. Poor maintenance of the air-conditioning ducts usually leads to the growth of fungus and molds inside the ducts.

In the category based on the nature of constituting chemicals, the chemicals used in paints, pesticides, furniture, thinners and solvents, detergents, and so on are essential constituents responsible for the chemical contamination of indoor air. The problem is compounded when there is closed, ineffective ventilation, thus leading to increased levels of contaminants for a prolonged time. Acids, any spillage of chemicals, aerosols, hydrocarbons, and more can also lead to the contamination of indoor air.

Further, the pollutants are classified based on the size of the particles. They may be in a state of liquid or solid, which, due to their light weight, can remain suspended in the indoor air and is thus liable to be inspired into the breathing air by the inhabitants of the affected building. The suspension of these particles can also occur due to the construction-related activities inside the buildings, cleaning activities, maintenance-related activities, and the activities related to the functioning of electronic equipment such as printers and photocopier.

1.4 Factors Affecting Indoor Air Quality

Following are the factors (Lee et al. 2002; Schieweck et al. 2018) affecting the quality of the indoor air: The movement of the air inside the building, ventilation provisions, regulation of temperature and humidity, effectiveness and efficiency of air conditioning, presence of particles suspended in air, emission of smoke, control of asbestos, gases released from the combustive activities inside the building, accumulation of radon occurring in long winter spells when the building is closed, ineffective ventilation of expired gases arising from the inhabitants of the building, and infestation with molds and microorganisms on account of poor sanitation and maintenance of the building.

Among these enumerated parameters, the humidity, regulation of temperature, and movement of the air are the most important ones. The overall perception of the quality of indoor air improves dramatically with the improvement of these critical physical parameters. Further, due to change in the humidity, temperature, and movement of air, the concentration of pollutants can increase or decrease drastically.

- **Temperature:** The temperature of the circulating air inside the building has a profound and determining effect on the perceived comfort of the inhabitants of the building. There are numerous other factors that influence the temperature levels of the indoor air: namely, effectiveness of the air-conditioning temperature control, sunlight entering into the building, lighting inside the building, equipment or electronic gadgets (such as computers, electronic appliances, water heaters) and humidity. Seasonal variations in temperature as per local weather and the ability of the installed temperature regulatory mechanisms have their effect on the temperature of the indoor air.

- **Humidity:** The effect of humidity on the comfort levels is mainly due to alteration exerted on the ability of the body to regulate temperature by perspiration. Whenever the humidity levels increase, the perspiration is not able to lower the body temperature and consequently, the perception of thermal heat will increase, thereby lowering the comfort level. Also, the high humidity levels in the indoor air augment the growth of molds, fungus, and other bacteria in the buildings. On the other hand, a decrease in humidity levels leads to the (a) drying of mucous membranes of the nose, mouth, and throat; (b) dryness in the eyes (causing pain); (c) itching in the affected areas, (d) infection in the affected areas. The low humidity levels can otherwise interfere with smooth operations of electronic gadgets.

- **Air Movement:** The circulation of the air around the occupants of the building has an effect on the perception of comfort for temperature. Also, it ensures the fast clearing of the pollutants and does not allow for the accumulation of them in the indoor air. Further, the circulation of air via carrying of heat along with air current regulates the humidity as well as the temperature of the indoor air. In case of increased temperature or humidity or both, good air circulation around the body is likely to increase the level of perception of thermal comfort. It is to be understood that the circulation of the air is also dependant on the flow of air current. The hot air becomes lighter and moves upward while the cold, dense air relatively settles in the lower parts of the rooms of the building. Wherever there will be obstructions to these air flow currents, the level of the thermal comfort is likely to go down.

- **Air Contaminants:** As such, the contaminants of the indoor air are from a vast range of substances—both living and nonliving. They have their origin in the objects used in the construction of the building, chemicals used in the maintenance of the building, physical parameters such as temperature and humidity leading to the growth of fungus and other microorganisms, furniture and other equipment installed in the building, as well as buildings' interaction with the external environment.

- **Respirable Suspended Particulates (PM10):** These are particles that are of a diameter less than or equal to 10 microns. Because of such size, these particles get suspended in the air whenever agitated through the passing of energy within them. Since they get suspended in the indoor air, they become liable to get inhaled through the breathed air, thus causing a harmful effect on the bodies of the affected persons. These particles can be of both living and nonliving types. Among the living ones, they may be the fungus, molds, and bacteria; pollen arising from plants; animal dander; insect origin particulate matter, and more. Whereas among the nonliving particulate matter, the most important ones are asbestos fibers; particulate matter arising due to combustive activities; tobacco; and other particulate matters arising due to smoking, cooking activities, and, in certain rare cases, radioactive particulates.

The effect of the particles largely depends on the shape, the size of the particulate, the density of particulate in the air, and the inertness of the particles. The particles with diameters of 2.5–10 microns are inhalable, whereas particles below the size of 2.5 microns come into the category of fine particles. The suspended particles emanating from tobacco smoking can go deep into the lungs, causing all the attendant problems within the respiratory system and, in case of prolonged exposures, can lead to carcinoma of the respiratory system. The particulates of biological origin can lead to allergic conditions, such as running of eyes and nose, asthma, and so on. The exposure to asbestos fibers arising due to construction activities creates a grave risk for restrictive lung diseases and cancer such as mesothelioma.

- **Volatile Organic Compounds:** These chemicals are of the hydrocarbon group, possessing carbon atoms and bearing a tendency for evaporation at even room temperatures. Hence, they are likely to enter the indoor air easily, in cases of contamination. More often, these gases do not bear color, thereby making their presence in the indoor air felt with even more difficulty. As of the writing of this chapter, about 900 such volatile compounds have been identified to be acting as pollutants of indoor area.

These chemicals can emanate from various construction materials, chemical agents used for cleaning and maintenance activities, pest control, cosmetic products, aerosols, thinners and paints, copying gadgets, writing ink, and more. Once a substantial level of concentration in the air indoors is achieved, their effect on the human body starts appearing.

The primary body systems more vulnerable to the adverse effect of these compounds are the nervous system, the blood, the hepatic system, and the urological system. Individual susceptibility to an allergy to such compounds can aggravate the allergy in affected individuals. Since the presence of such compounds may mostly go unnoticed, long-term exposures may occur even without those affected knowing. Chronic exposure to low doses may also cause reactions. The benzene present in such compounds has been found to cause cancers in cases of long-term exposures to them.

In a definition given by The International Labour Organization (ILO), "Sick Building Syndrome" has been defined to be the presentation of symptoms by about 20% of inhabitants of a building wherein a relation could be shown between the conditions of the building and connection with the quality of indoor air.

1.5 Materials and Methods

- **Study Design:** The present study is a descriptive study that is aimed to collect big data regarding indoor air quality from a randomly selected corporate office or data available in public domain from the Internet, and to apply suitable compression technique among the available techniques to the big data so obtained, thereby learning and understanding the shortcomings and strengths of data compression techniques.

- **Place of Study:** The study has been done in the state of Delhi. A corporate office was selected by purposive sampling from the list of available corporate offices in the state of Delhi. However, due to the time constraint, the permission could not be obtained, and the big data available in public domain on the Internet was obtained and the suitable data compression technique was applied on the data so collected.

- **Inclusion Criterion:** The corporate office available in the state of Delhi that had more than five floors in the building was selected.

- **Sampling Technique:** Purposive sampling method was used.

- **Data Collection:** Secondary data was collected from the Internet for this study; due to the time constraint, the permission could not be obtained from the purposively selected corporate office.

- **Data Analysis:** The data was compressed with a suitable data compression technique selected from among the most commonly available data compression techniques.

- **Study Limitations:** The study findings are for the selected data only. The extrapolation of the same may or may not yield the same results.

- **Ethical Considerations:** As the data was used as available in the public domain on the Internet and no human subjects were involved, the study did not attract ethical considerations as such.

- **Data Collection for Monitoring of Indoor Air Quality:** Several data were collected inside the building for monitoring and surveillance of indoor air quality so that any deviation from the desired values might be taken notice of as soon as possible, and corrective measures could be instituted quickly enough to avoid any harmful consequence on the occupants of the building. The parameters collected are temperature, humidity levels, parameters related to air circulation and flow, levels of various gases and volatile organic compounds, levels of radon, and so on (Lee et al. 2002).

  The data so collected were used for the assessment. The assessment can be made in two ways based on the collection mode of data, such as the data collected on a real-time basis with the continuous mode, or data collected to be coupled with the analysis (both the computer analysis and the laboratory analysis). In the case of monitoring being done in real time, the corrective actions may be taken early and promptly. Further, to have active monitoring take place, the minimum number of sampling points in a building is to be monitored. For example, for a building higher than or equal to 30,000 square meters, the minimum number of sampling points taken are 1 per 1200 meters. The need for any further sampling points may be carried out by the size and type of the building.

The time frames may usually be taken worldwide for such monitoring and data collection for the quality of indoor air taken at an interval of eight hours. In cases of practical or technical difficulties faced in a continuous real-time collection, data may be collected by using the reading taken for 30 minutes made 4 times in a day. However, it is to take care that this data capturing is adequately done during the working hours. It is better to capture data when the footfall of public in the building is highest. In case of using monitors with real-time data capturing ability, an interval of 5 minutes for each point of collection of data should be done. Such data may be fed to computers for continuous analyzing and application of information so arrived for rectification, if required.

- **Challenge of Massive Data Management:** It may be appreciated that collecting real-time data from the sampling points in a building with the frequency of 5 minutes creates a large data (over time) that is more challenging. As such data collected for several variables, several times a day around the year and ongoing creates a massive amount of data that results in sorting, handling and storing challenges.

The data so collected from different sampling sources, mostly collected through automatic smart mode, since is voluminous in size, is referred to as "big data." More often than not, the vast data so collected is not clean and contains unnecessary, redundant data that increases the cost to the company by using space at a substantial cost.

The following are the significant challenges faced:

1. The data is vast; however, the same is to be stored for the regulatory compliances even if the data is no more meaningful in the current scenario.

2. The data is managed, analyzed, and collated by the Information Technology Department, which usually has an insufficient budget. To serve the duty of management of such massive data within the allocated budget is a challenge.

3. The accurate and complete retrieval is a challenging task since the data may get altered or fail to be retrieved.

4. The cost of storage is a deterrent to the safekeeping of such enormous data.

5. Another challenge is the variety of the data. In the context of the air quality, several parameters in real time have to be collected which may be raw, structured, semi-structured, or unstructured data. The complicated relationship between these data types cannot be efficiently processed by the traditional ways of storage (Kadadi et al. 2014).

6. Such big data and its management entail velocity, which deals with the speed at which data is received and transmitted from various devices. Especially the real-time data collection, processing, analysis, and following application of the information on management of air quality poses a critical challenge.

7. Lastly, the big data of air quality monitoring is complex and diverse, for active management, connection, and the correlation between hierarchies of data.

Compression of data is a means of management of big data arising in indoor air quality monitoring per the challenges described in the preceding list. It makes sense to compress the data to save the cost, to ensure the safe storage of data for regulatory compliances, and to ensure accurate and complete future retrieval. The data compression techniques employed for such compression of data are as follows: For text data, arithmetic coding, Huffman Algorithm, LZW coding, run length encoding techniques, and so on may be used. It is to be understood that the choice of compression technique is dependent on the type of data files, whether they are text files, image files, audio files, or video files. For data consisting of text or programs, lossless techniques are employed, whereas for data consisting of images, audio files, or video files, lossy techniques are used for data compression. The techniques covered under lossless techniques are run length encoding, entropy encoding, and adaptive dictionary algorithms such as Lempel-Ziv-Welch encoding. On the other hand, the techniques employed under the lossy category are JPEG coding for images, MPEG coding for videos, and predictive encoding and perceptual encoding techniques (Bonde and Barahate 2017).

1.6 Big Data Management System

The compression of the data is a solution in terms of the volume of the data. However, the compressed data is also to be managed. The cloud computing stands out as a solution for its management. It allows data processing over thousands of servers, thereby reducing the cost and providing agility to the processing of the compressed data.

Further, it provides feasibility to the compressed data management as it removes the need for the physical servers and performs the same function, albeit by the virtual servers. Cloud computing for big data management is done through analytics as a service method. It can be made to work through services using software, platform, or infrastructure mode. However, for analytics as a service method to work effectively, it must contain an address for six elements:. data sources, data models, processing applications, computing power, analytic models, and sharing of outcomes of analysis made. Thus the essential ingredient of cloud computing is enabling the web to be used as a space for these processes (Kumar and Goudar 2012; Saxena et al. 2012).

- **Software as a Service (SaaS):** In this method, the ready-made applications are used to work on a network. There is an option of using multiple ready-made applications at the same time. The disadvantage experienced in this method is that this does not allow for any alteration to be made. However, the advantage is that this method has fewer compatibility issues and is easy to use. This is a cost-effective method also and can be used by small-scale companies. Also, updating is facilitated automatically.

  Examples of SaaS include SalesForce.com, Google Mail, Google Docs, and more.

- **Platform as a Service (PaaS):** This is a method whereby virtual servers can be created to run several applications simultaneously. A significant advantage offered in this method is that it can be used by the people located at places over substantial geographical distances. It can also be upgraded easily.

  An example of a platform-based service method is Google AppEngine.

- **Infrastructure as a Service (IaaS):** In this type of method, the cloud service is obtained by another provider. The analytics software is installed by the recipient organization. Its main advantage is that the tools required for efficient analysis are provided by the service provider, thus making the computing faster. This type of method due to faster computing is more suited when different sets of data are to be analyzed quickly.

An example of such a kind of service is Amazon Web Services (Shrivastava and Bhatnagar 2011; Sharma and Bhatnagar 2011; Ashraf 2014).

1.7 Big Data Processing Using Distributed Data Storage

A key challenge experienced in the processing of big data is to access several application servers, at the same time thus generating a sizeable number of read as well as write requests. Such read and write requests for the effective access of data require several servers to be taken in position for storage of data so that a proper distribution of reading/writing requests over several such servers is made to avoid the interruption of the whole service if a substantial number of servers get failed (Tsuchiya et al. 2012; Elzein et al. 2018).

Distributed key-value store is a type of technique in which the processing of data is made through the distribution of a data store. In this method, the structure of data consists of key and value stored in a distributed manner over several servers, whereas reading and write commands happen in another server by a specific request made by a key in return. Thus the advantage comes from avoiding accumulation of read/write requests on a single server, thereby increasing the overall performance. This feature also reduces the events and risks of failure of servers as the copying and storing of data are being made on other servers, thereby ensuring that if one server fails, the data is not lost. This type of technique is well suited for the websites where several users usually access the website at one single point of time, for instance, online shopping sites (Tsuchiya et al. 2012).

1.7.1 Algorithm Process for Data Compression Using the Specified Techniques

In this article, big data are collected from the internet from the URL https://catalog.data.gov/dataset/air-quality-ef520 maintained by the US government; the data collected is in CSV format (Comma Separated Value file format). To understand, learn, and apply the identified data compression technique, Snappy, previously known as Zippy (Jägemar et al. 2016) is applied to the big data collected and then compared with another compression technique, Optimized Row Columnar (ORC). The following steps are followed:

1. The big data source file is obtained from the public domain (in this case, data.gov website).
2. The file is placed in the source directory of the local file system (in this case, Ubuntu operating system is used).
3. Snappy is chosen and applied as the suitable compression technique among the available compression techniques to compress the given data.
4. Hadoop Shell and Hive Shell commands are used for compressing (using Snappy as the compression technique) followed by the transfer of the compressed file to Hadoop File System (HDFS).
5. After the completion of the compression process, the output compressed file in the Hadoop File System is analyzed and compared with the uncompressed input file for space and time parameters.

The following commands are used for data compression using Snappy:

1. LAUNCH HIVE SHELL USING $ cd $HIVE_HOME and $ bin/hive. ENTER INTO HIVE SHELL
2. CREATE EXTERNAL TABLE IF NOT EXISTS Air_Quality(indicator_data_id INT,indicator_idINT, nameSTRING, MeasureSTRING, geo_type_nameSTRING, geo_entity_idINT, geo_entity_nameSTRING, year_descriptionINT, data_valuemessage FLOAT) ROW FORMAT DELIMITED FIELDS TERMINATED BY ',' STORED AS TEXTFILE;
3. LOAD DATA LOCAL INPATH '/home/khushboo/Desktop/Air_Quality.csv' OVERWRITE INTO TABLE Air_Quality;
4. CREATE EXTERNAL TABLE IF NOT EXISTS Air_Quality_new(indicator_data_id INT,indicator_idINT, nameSTRING, MeasureSTRING, geo_type_nameSTRING, geo_entity_idINT, geo_entity_nameSTRING, year_descriptionINT, data_valuemessage FLOAT) STORED AS ORC LOCATION '/user/hive/orc' TBLPROPERTIES ("orc.compress"="SNAPPY");
5. INSERT OVERWRITE TABLE Air_Quality_new SELECT * FROM Air_Quality;

After the application of Snappy technique, Optimized Row Columnar (Jukic et al. 2017) is applied to the same input file. The ORC technique mainly refers to stimulating the employment of analytical databases. It integrates columnar method with a technique based on denormalizing the data tables for analysis and decision support. This technique is not to be discussed in detail in this chapter, as it is used only for comparison.

1.8 Results

First, big data has been collected regarding indoor air quality from a randomly selected corporate office or data available in the public domain from the Internet. Then, a compression technique called Snappy has been applied as mentioned in the algorithm in Section 1.7.1.

Figure 1.1 shows the screenshot of the launch of hive shell launched using the commands mentioned in the algorithm, and Figure 1.2 shows the screenshot of the creation of the table (named Big Data) having loaded file data. Figure 1.3 indicates the application of Snappy compression technique.

The same input file used in Snappy compression technique is used in ORC technique after that. Figure 1.4 shows the loading of the Air Quality input data file into t1 directory of the Hadoop file and launching of the hive shell. Figure 1.5 shows the creation of a table t2 and loading into it the input file

FIGURE 1.1
Hive shell launched.

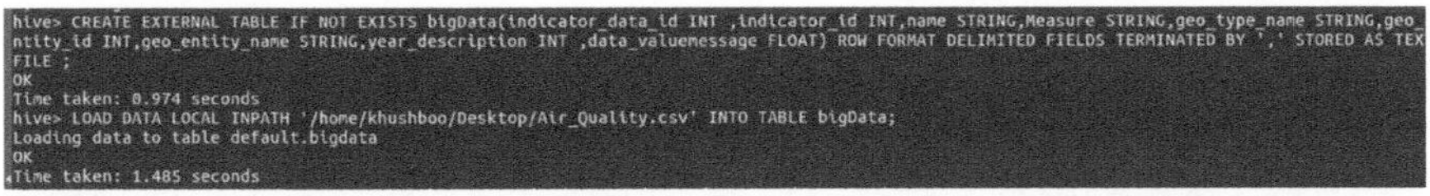

FIGURE 1.2
Creation of "Big Data" table.

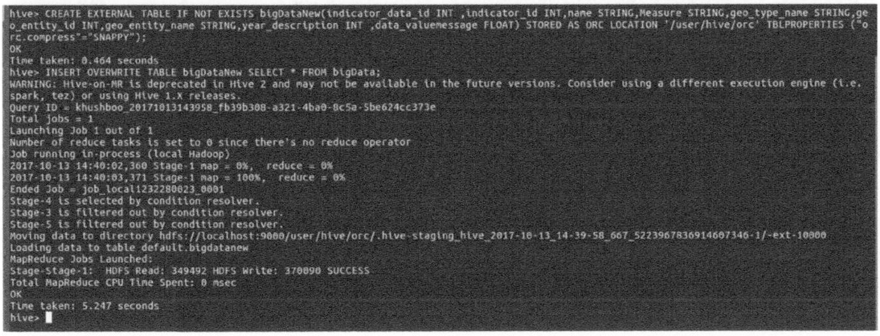

FIGURE 1.3
Application of Snappy compression technique.

data followed by the loading of the compressed file in ORC format into table t3 that is stored into the Hadoop file system.

Figure 1.6 shows the screenshot of the actual uncompressed input file used for the compression. Its size is 349.5 kB. Figure 1.7 shows the compressed

FIGURE 1.4
Loading input data file into HDFS.

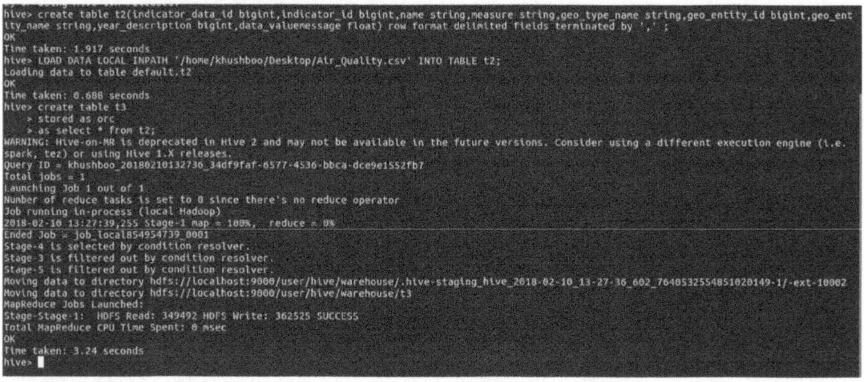

FIGURE 1.5
Creation and loading of table t2 with the input file data as well as loading of compressed ORC file into table t3.

file using the Snappy technique, while Figure 1.8 shows the same input file compressed by ORC.

After comparing Snappy with ORC technique, it may be seen from Table 1.1 that in the Snappy technique, the time taken for compression has been found to be 0.464 seconds, whereas in ORC technique, the time taken for compressing the same input file has been found to be 3.24 seconds: that is, Snappy technique is approximately seven times faster than ORC technique, but the former one gives somewhat less compression ratio than the latter one

FIGURE 1.6
Uncompressed input file.

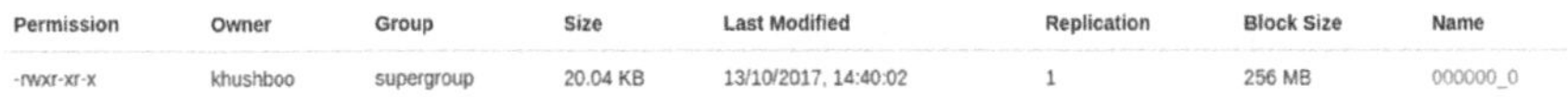

Permission	Owner	Group	Size	Last Modified	Replication	Block Size	Name
-rwxr-xr-x	khushboo	supergroup	20.04 KB	13/10/2017, 14:40:02	1	256 MB	000000_0

FIGURE 1.7
Compressed input file (using snappy).

Permission	Owner	Group	Size	Last Modified	Replication	Block Size	Name
-rwxrwxr-x	khushboo	supergroup	12.66 KB	10/02/2018, 13:27:38	1	256 MB	000000_0

FIGURE 1.8
Compressed input file (using ORC).

TABLE 1.1

Comparison Between Snappy and ORC Compression Techniques

Compression Technique	Time Taken (sec)	Original Size (A)	Compressed Size (B)	Compression Ratio (C = A/B)
Snappy	0.464	349.5 kb	20.04 Kb	17.44
ORC	3.24	349.5 kb	12.66 Kb	27.61

TABLE 1.2

Comparison Between Snappy, LZW, LZO, and LZMA Compression Techniques

Compression Technique	Compression Speed	Compression Ratio
Snappy	Very fast	Medium
Lempel–Ziv–Welch (LZW)	Medium	Medium
Lempel–Ziv–Oberhumer (LZO)	Fast	Low
Lempel–Ziv–Markov chain algorithm (LZMA)	Slow	High

(although both Snappy and ORC techniques yield the compression ratio in the medium range).

Because of the very high speed of Snappy, it has been preferred in the present research over ORC. A very brief comparison of Snappy with some other existing techniques is also shown in Table 1.2.

By observing the different types of compression techniques, it may be concluded that Snappy gives the best overall performance over the other three compression techniques.

1.9 Conclusion and Future Scope

The primary motive of this research study was to investigate and analyze indoor air quality and compress the big data file in CSV format using a suitable compression technique. The data file was collected from the public domain data.gov. Through this study, understanding and application of primarily Snappy technique and secondarily ORC technique have been discussed to compress the given data file. There is an essential requirement to compress the big data to save the cost, ensuring safe storage and convenient management of data.

A small comparison has been drawn between the Snappy and other compression techniques. Snappy has been chosen as the suitable compression technique as it is considerably faster as compared to other mentioned data compression techniques like ORC, LZW, LZO, and LZMA. The compression ratio obtained is also a good one (although not too high but also not too low).

The analysis of indoor air quality and its compression have a lot of practical implications, as it has been experienced worldwide that there is a magnitudinous increase in data volume. It is estimated that the data of an organization in our modern world is almost doubling every year, thereby creating a massive challenge of storing this big data. So, it becomes mandatory to reduce the size of the data, and the practical implications of compression are for the effective managing, storing, processing, and analyzing the big data.

It is also suggested that the compression being executed at the data resources may be further implicated in the future to find the actual on-site time taken for compression as well as storage of such voluminous big data about the parameters of indoor air quality. The only limitation of the present research is that the study findings are for the selected data only. The extrapolation of the same may or may not yield the same results. The purposive sampling for the corporate office, however, faced with the constraint of the time required for permission for data collection, could not be obtained, and the data available in the public domain has been used.

References

Alker, John, Michelle Malanca, C. Pottage, and R. O'Brien. "Health, wellbeing & productivity in offices: The next chapter for green building." *World Green Building Council* 2014.

Apter, Andrea, Anne Bracker, Michael Hodgson, James Sidman, and Wing-Yan Leung. "Epidemiology of the sick building syndrome." *Journal of Allergy and Clinical Immunology* 94, no. 2 (1994): 277–288.

Ashraf, Imran. "An overview of service models of cloud computing." *International Journal of Multidisciplinary and Current Research* 2, no. 1 (2014): 779–783.

Bonde, Poonam and Sachin Barahate. "Data compression techniques for big data." *Journal of Emerging Technologies and Innovative Research* 4, no. 3 (2017).

Cociorva, Sorin and Andreea Iftene. "Indoor air quality evaluation in intelligent building." *Energy Procedia* 112 (2017): 261–268.

Dhall, Rohit and Vijender Kumar Solanki. "An IoT based predictive connected car maintenance approach." *International Journal of Interactive Multimedia and Artificial Intelligence*, 2017.

Elzein, Nahla Mohammad, Mazlina Abdul Majid, Ibrahim Abaker Targio Hashem, Ibrar Yaqoob, Fadele Ayotunde Alaba, and Muhammad Imran. "Managing big RDF data in clouds: Challenges, opportunities, and solutions." *Sustainable Cities and Society* 39 (2018): 375–386.

Heinsohn, Robert Jennings and John M. Cimbala. *Indoor Air Quality Engineering: Environmental Health and Control of Indoor Pollutants*. CRC Press, Boca Raton, FL, 2003.

Heseltine, Elisabeth and Jerome Rosen, eds. *WHO Guidelines for Indoor Air Quality: Dampness and Mould*. WHO Regional Office Europe, 2009.

I.A.Q. Determine, Housekeeping Plans. "IAQ building education and assessment model." 2008.

Jägemar, Marcus, Sigrid Eldh, Andreas Ermedahl, and Björn Lisper. "Automatic message compression with overload protection." *Journal of Systems and Software* 121 (2016): 209–222.

Jansz, Janis. "Theories and knowledge about sick building syndrome." In *Sick Building Syndrome*, pp. 25–58. Springer, Berlin, Germany, 2011.

Jones, Andy P. "Indoor air quality and health." *Atmospheric Environment* 33, no. 28 (1999): 4535–4564.

Jukic, Nenad, Boris Jukic, Abhishek Sharma, Svetlozar Nestorov, and Benjamin Korallus Arnold. "Expediting analytical databases with columnar approach." *Decision Support Systems* 95 (2017): 61–81.

Kadadi, Anirudh, Rajeev Agrawal, Christopher Nyamful, and Rahman Atiq. "Challenges of data integration and interoperability in big data." In *Big Data (Big Data), 2014 IEEE International Conference on*, pp. 38–40. IEEE, 2014.

Kadam, Vrushali, Sharvari Tamane, and Vijender Solanki. "Smart and connected cities through technologies." In *Big Data Analytics for Smart and Connected Cities*, pp. 1–24. IGI Global, Hershey, PA, 2019.

Kephalopoulos, Stylianos, Dimitrios Kotzias, and Kimmo Koistinen. European collaborative action, Urban air, indoor environment and human exposure, "Impact of ozone-initiated terpene chemistry on indoor air quality and human health." Report No. 26, 2007.

Kumar, Santosh and R. H. Goudar. "Cloud computing-research issues, challenges, architecture, platforms and applications: A survey." *International Journal of Future Computer and Communication* 1, no. 4 (2012): 356.

Lee, Shun-Cheng, Hai Guo, Wai-Ming Li, and Lo-Yin Chan. "Inter-comparison of air pollutant concentrations in different indoor environments in Hong Kong." *Atmospheric Environment* 36, no. 12 (2002): 1929–1940.

Lister, Debra Brinegar, Elisabeth M. Jenicek, and Paul Frederick Preissner. *Productivity and Indoor Environmental Conditions Research: An Annotated Bibliography for Facility Engineers*. DIANE Publishing, 1998.

Maitrey, Seema and C. K. Jha. "MapReduce: Simplified data analysis of big data." *Procedia Computer Science* 57 (2015): 563–571.

Nathanson, Tedd. *Indoor Air Quality in Office Buildings: A Technical Guide: A Report of the Federal-Provincial Advisory Committee on Environmental and Occupational Health*. Vol. 65. Canadian Government Publishing, 1993.

NEMI (National Energy Management Institute). *Productivity Benefits Due to Improved IndoorAir Quality*. NEMI, Alexandria, VA, 1994.

Occupational Safety and Health Administration. *Indoor Air Quality in Commercial and Institutional Buildings*. Maroon Ebooks, 2015.

Persily, Andrew K. "Evaluating building IAQ and ventilation with indoor carbon dioxide." *Transactions-American Society of Heating Refrigerating and Air Conditioning Engineers* 103 (1997): 193–204.

Roulet, Claude-Alain, Flavio Foradini, Claude-A. Bernhard, Lucio Carlucci, and JOULE II. "European audit project to optimise indoor air quality and energy consumption in office buildings." *National Report of Switzerland* (1994): 1–60.

Sahlberg, Bo. "Indoor environment in dwellings and sick building syndrome (SBS): Longitudinal studies." PhD dissertation, Acta Universitatis Upsaliensis, 2012.

Sanju, Durga Devi, A. Subramani, and Vijendar Kumar Solanki. "Smart city: IoT based prototype for parking monitoring & parking management system commanded by mobile app." In *Second International Conference on Research in Intelligent and Computing in Engineering*, 2017.

Saxena, Agreeka, Gulshan Shrivastava, and Kavita Sharma. "Forensic investigation in cloud computing environment." *The International Journal of Forensic Computer Science* 2 (2012): 64–74.

Schieweck, Alexandra, Erik Uhde, Tunga Salthammer, Lea C. Salthammer, Lidia Morawska, Mandana Mazaheri, and Prashant Kumar. "Smart homes and the control of indoor air quality." *Renewable and Sustainable Energy Reviews* 94 (2018): 705–718.

Sharma, Kavita and Vishal Bhatnagar. "Private and secure hyperlink navigability assessment in web mining information system." *International Journal on Computer Science and Engineering* 3, no. 6 (2011): 2245–2250.

Shrivastava, Gulshan and Vishal Bhatnagar. "Secure association rule mining for distributed level hierarchy in web." *International Journal on Computer Science and Engineering* 3, no. 6 (2011): 2240–2244.

Solanki, Vijender Kumar, M. Venkatesan, and S. Katiyar. "Conceptual model for smart cities for irrigation and highway lamps using IoT." *International Journal of Interactive Multimedia and Artificial Intelligence*, 2018a.

Solanki, Vijender Kumar, M. Venkatesan, and S. Katiyar. "Think home: A smart home as digital ecosystem in circuits and systems," Vol. 10, No. 7, Scientific Research Publishing, 2018b.

Solanki, Vijender Kumar, Somesh Katiyar, Vijay Bhaskar Semwal, Poorva Dewan, M. Venkatesan, and Nilanjan Dey. "Advance automated module for smart and secure city." In *ICISP-15, G.H. Raisoni College of Engineering & Information Technology*, Nagpur, India, December 11–12, 2015, Procedia Computer Science, Elsevier.

Spengler, John D., Jonathan M. Samet, and John F. McCarthy. *Indoor Air Quality Handbook*, 2001.

Tsuchiya, Satoshi, Yoshinori Sakamoto, Yuichi Tsuchimoto, and Vivian Lee. "Big data processing in cloud environments." *FUJITSU Science and Technology Journal* 48, no. 2 (2012): 159–168.

World Health Organization (WHO). "Global health risks—mortality and burden of disease attributable to selected major risks." *Cancer* (2017).

Yang, Chi, Xuyun Zhang, Changmin Zhong, Chang Liu, Jian Pei, Kotagiri Ramamohanarao, and Jinjun Chen. "A spatiotemporal compression based approach for efficient big data processing on cloud." *Journal of Computer and System Sciences* 80, no. 8 (2014): 1563–1583.

2

Programming Paradigms for IoT Applications: An Exploratory Study

Sivadi Balakrishna and M. Thirumaran

CONTENTS

2.1 Introduction

Internet of Things (IoT) is used as a communication medium for things and the Internet. The Internet of Things–based use cases, case studies and frameworks are applied on the sectors of healthcare, industry 4.0, logistics 4.0, smart city, smart energy, smart grid, smart building, smart transport management, smart retail, and more. These all are fully operational, customer management, business supportable, and scalable. The goals included for these use cases based on the Internet of Things are better customer service, enriching the disruptive data source, tortious intelligence, and innovative approach. This chapter is totally focused on various supporting programming technologies for IoT applications in an exploratory case study way. Overall, five case studies have been studied to fill the gaps between the researchers' sides to IoT platforms' side.

Singh D et al. [1] mentioned that the Internet of Things is an emerging technology for communicating through the Internet and communication technologies. In simple words, we can say that the "Internet of Things" is nothing but connecting living and nonliving things to the Internet. Billions of heterogeneous resources are connected to the Internet, not only from sensors and actuators but also from various IoT deployment models, a huge variety of data, a high volume of data and low-level descriptive resources. Since 2010 researchers have been analyzing and implementing the IoT-based smart city and industry 4.0 applications by applying various kind of frameworks and approaches. To make the service universally maintainable, web accessibile and accurately open, the IoT combines the web and the Internet of Things, which emerges as the web of Things (WoT). Smart objects have been treated as everything in the IoT platform, and these smart objects allow communication with each other through the Internet physically or virtually. Per S. López et al. [2], the IoT helps for connecting people or things at anytime, anywhere, and with anything using a network path or service. There are also still primary resources on the Internet that require more attention to be able to access those things' or resources' interaction and monitoring by applications and interfaces. Sivadi Balakrishna et al. [3] estimate that there will be more than 50 billion connected devices to the web by 2020, and that represents a fivefold increase from 10 billion in 2010.

The Internet of Things (IoT) is used to connect things to the Internet and is a combination of IoT devices: sensors, actuators, Radio Frequency Identification (RFID) tags, and smoothly distributed smart IoT objects having sensing and actuating capabilities, embedding with IoT technology [4]. The IoT mainly addresses scalability, accessibility, visibility and controllability of the sensing smart objects and things. In the future, the physical objects and digital objects have to be embedded and intercommunicated to obtain more domain-specific applications [5]. The IoT concentrates on transforming the real-time objects into sensible smart objects with communicative and controllable environmental physical objects. RFID is the technology used to capture objects, people and living and nonliving things [6]. Electronic Product Codes (EPC) are embedded RFID tags to be used for tracking IoT smart things. The cloud and big data technologies are the finest technologies that are useful for storage and performing analysis on IoT data [7]. The IoT has a mid-range list of applications to be supportable and suitable for smart city environments: healthcare applications [8], environmental monitoring [9], supply chain management of food [10], production and inventory management [11], fire station systems [12], Parking management data [13], connected cars [14], highway lamps [15], smart homes, smart and secure cities [16], social networks and semantic real-time traffic management [17]. Vijender Kumar Solanki et al. [18] concluded that the smart home requires appropriate data for maintaining smart city applications that are more convenient and useful to the users.

The first case study of this chapter is on ThingSpeak—an IoT cloud platform for IoT analytics. This is the medium to collect and analyze the data, visualize the collected data within a MATLAB environment and finally "act" the data, meaning to react to data. ThingSpeak mainly consists of channels, apps, and the community. Here everyone has its own role of placement for users. This case study demonstrates how to post the collected data from external sources to a ThingSpeak channel. For that use the ESP8266 board followed by the ThingSpeak-IoT cloud platform for both posting data as well as retrieving data from a created channel.

The second case study is fully focused on the sending of sensor data to the Ubidots cloud platform using Wi-Fi through RESTful API: taking the sensor data coming from various IoT devices like sensors, actuators, global positioning systems (GPSs), laser scanners and RFID cards and sending it to Ubidots—a cloud platform for online analysis and visualizations.

The third case study is on Arduino, for connecting tiny IoT devices. Arduino is an open source hardware for connecting tiny IoT devices to the Internet. It was developed by Massimo Banzi and David Cuartielles in 2005. Coding is accessible and transferable in all languages like C++, Java, and Python.

In fourth case study deals with LinkIt Smart AWS (Amazon web services) IoT. This case study shows IoT developers how to use the Amazon web services and IoT applications step by step, in an experimental way. These are the following steps that illustrate the instructions for preparing the LinkIt Smart AWS IoT.

Step 1: Create a Things in AWS IoT and generate and download AWS IoT device certificates.

Step 2: Create and attach a policy to the generated certificate.

Step 3: Connect LinkIt Smart to the Internet.

Step 4: Transfer the AWS IoT device certificates from your computer to LinkIt Smart using Winscp.

Step 5: Run the code.

Step 6: Create a rule for the AWS IoT Thing and upload the data points from LinkIt Smart to DynamoDb.

The last case study studies the RESTful CoAP application layer protocol for tiny IoT-based applications. CoAP protocol consumes less energy for getting IoT resources. R.T. Fielding [19] argued that the term "web services" plays a crucial role in developing Internet-based applications. If one service in one web application needs to talk with another service in another application, that is possible with REST architecture. Using REST, each device can be easily be made available to state its information and its strength in terms of scalability, statelessness, cacheability, interoperability, and efficiency. It permits simple incorporation with various web browsers and diverse service providers. The main objective of the RESTful-based CoAP protocol is that it can easily translate to HTTP for integration of constrained networks with the World Wide Web. In this case study is presented a new and emerging CoAP Protocol in IoT technology that makes it possible to integrate things easier. It is designed for small devices with low-power sensors and actuators for managing things over the Internet through standard network properties. In real-time applications like U-healthcare systems, the RESTful CoAP protocol is used to receive the sensing data and producing local analysis, which finally it will send to the Remote Medical Server (RMS) for analysis.

The next sections of this chapter are organized as follows: Section 2.2 shows the case study on ThingSpeak—an open IoT Cloud platform. The second case study fully focuses on sending sensor data to the Ubidots cloud platform using Wi-Fi through RESTful API in Section 2.3. In Section 2.4 is shown the case study on Arduino for connecting tiny IoT devices. Section 2.5 demonstrates the LinkIt Smart AWS IoT applications. The last case study studies the RESTful CoAP application layer protocol for tiny IoT-based applications as depicted in section 2.6. Finally, Section 2.7 concludes this chapter.

2.2 A Case Study on ThingSpeak—An Open IoT Cloud Platform

This section describes ThingSpeak—an IoT Cloud platform (available at https://thingspeak.com). This is the medium to collect the data, analyze, and visualize the collected data with MATLAB environment and finally "act" the data, meaning to react to data. It mainly consists of channels, apps, and community. Here every one has its own role of placement for users. This case study demonstrates how to post the collected data from external sources to a ThingSpeak channel. For that use the ESP8266 board followed by the ThingSpeak-IoT cloud platform for both posting data as well as retrieving data from a created channel.

2.2.1 Prerequisites

2.2.1.1 Supported Hardware

1. ESP8266-12
2. ESP8266-12E

The following are the list of steps to be followed for demonstrating post-temperature data and readings from the ThingSpeak channel.

1. Firstly, create a ThingSpeak channel; if you already have an account, just log in to it.
2. In the channel creation, make sure that at least two fields are enabled:one for reading data and one for writing data.
3. Record the Read API key and Write API key. These two types of keys are available in the API's keys tab from the channel stings view.
4. Using a browser address bar, all the required fields can be filled. Here some values are required as input for estimation of the temperature coefficient values.

2.2.1.2 Required Hardware

The ESP8266 hardware board used the Node MCU for development programs. The ADC pin should not expose on ESP8266-01. For that, use a digital interface sensor along with ESP8266-01.

- 10 KΩ resistor
- 10 KΩ transistor
- Breadboard
- Jumper wires (minimum 3)

2.2.2 Schematic and Connections

Before going to write code for the program, first, check the hardware connections. The following Figure 2.1 depicts Node MCU breadboard with a schematic connection.

- On Node MCU, connect the thermistor one pin to the pin A0.
- Next, the thermistor second pin is connected to the 10 KΩ resistors.
- The ground is connected by the second resistors. Make sure the ground connection is common to the Node MCU.

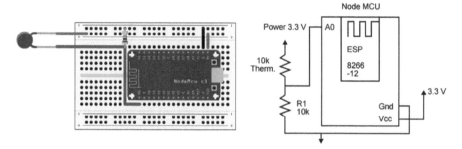

FIGURE 2.1
Node MCU board and schematic connection.

2.2.3 Program for ESP8266

The following list of steps is carried out to write the program on ESP8266.

1. First, download the latest Arduino IDE (available at https://www.arduino.cc).
2. Then, add the supporting ThingSpeak library for ESP8266 and Arduino.
    a. Select Sketch—Include library—Manage libraries.
    b. Select ThingSpeak by chosen sketch.
3. Connect the board package for ESP8266.
    a. File- Preference- Enter http://Arduino.esp8266.com/stable/package_esp8266com_index.json
    b. Goto Tools—Select Boards—again select Board Manager—Enter ESP8266 in the search bar and click Package install.
4. Select the corresponding board and port in the Arduino IDE. Node MCU 2.0 in the hardware is used to generate this.
5. Finally, create the application using these following steps:
    a. Open the new window of Arduino IDE and then save the file. Write the code in that window. Choose the Channel id—Read API key—Write API key—no need to change the code unless the application reads data from that channel.
    b. Once everything is connected, the device may find the voltage values every two minutes. It also measures the temperature values across the stored channel; after that it posts the temperature values to the channel as shown in Figure 2.2.

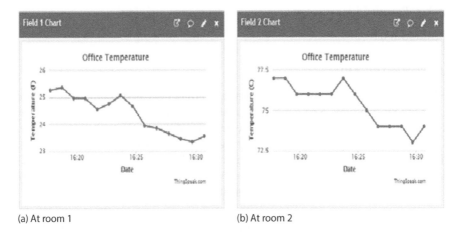

(a) At room 1   (b) At room 2

FIGURE 2.2
Temperature values at ThingSpeak.

2.3 A Case Study on Sending Sensor Data to Ubidots Cloud Platform Using Wi-Fi

This is the case study on using the Ubidots cloud platform to receive sensor data from various IoT devices using Wi-Fi. Here, the sensor data coming from various IoT devices like sensors, actuators, global positioning systems (GPS), laser scanners and RFID cards is sent to Ubidots—a cloud platform for analysis and visualizations at online.

Send sensor data to the cloud using Wi-Fi (over HTTP using RESTful API). The following are the steps for this case study:

1. First, create an account in Ubidots.
2. Then connect LinkIT Smart to the Internet with Wi-Fi.
3. Post sensor data to the cloud as JSON.
4. Create the plot and graph for visualization.
5. Create the triggering rule or event in Ubidots.

2.3.1 Required Software

Arduino IDE (version 1.6.6) and LinkIt ONE SDK, as given in installation guide.

2.3.2 Required Hardware

LinkIt Smart

2.3.3 RESTful API

REST defines simply the way of exchange of data between clients and servers over HTTP protocol. REST has the greatest principles like cacheable mechanism, statelessness, and uniform interface. Like the REST architecture style, CoAP is also using URIs for identifying resources through representations. These URIs can be treated as nouns and HTTP methods like GET, PUT, UPDATE, and DELETE treated as verbs. REST is an architectural style for connecting distributed hypermedia systems.

- **GET:** This command is used to get the information or data from the cloud.
- **POST:** This command is used to create a new thing inside the cloud.
- **PUT:** This command is used to update the already existed things.
- **DELETE:** This command is used to delete or remove the already existed things.

2.3.4 Ubidots IoT Cloud Platform

- Creating account and setup for variable.
- Go to the following address bar and create the new account: https:// app.ubidots.com/account/signup
- Log in to created account.
- Add data source in your account.
- Click on Sources, then click on Add Data Source. It will create My Data Source.
- Go to my data source and add variables to your data source.

Before sending sensor data to the Ubidots cloud, we will need to tell the API ID of the variable we want to update, and include a security token so that our request can be authenticated.

- **Variable ID:** This is a unique identifier of the time series that stores your data.
- **Token:** A unique key to authenticate your request.
  - Get a variable ID and Token.
  - Click top right corner on your < name > API credentials and copy token.

- Now go to Source > My Data Source < your variable > copy the ID (Variable ID).
- Creating plot and graph for visualization
  - Go to the Dashboard tab and click on Add widget.
  - Select the chart > Line chart > Add variable, and the variable you just created.
  - Click on Finish.
- Creating some triggering rule (event) in Ubidots
  - Send yourself an SMS.
  - Now go to the Events tab and choose Add Event.
  - Select the data source and the variable you just created.
  - Configure a condition to triggering rule, if the value of variable is > 100.
  - Click on Send SMS, then type mobile number with the country code and click on the Finish button.
  - You should now get an SMS alerting you, if the sensor value is above the given threshold!
- Storing Python program to LinkIt Smart open WRT Linux using WinSCP
  - Log in to LinkIt Smart thought serial USB to get the IP address of the LinkIt Smart, which is assigned by the Wi-Fi router/ mobile hotspot.
  - Use ifconfig command to get the IP address as follows:

```
#ifconfig
```

- This command will give you IP address as highlighted in the preceding screen shot (i.e., apcli0, Inet addr: 192.168.10.104).
- Your board's IP address may be different than this IP.
- Open WinSCP and click on New session.
- Select the file protocol as SCP, as shown in the following screen shot:

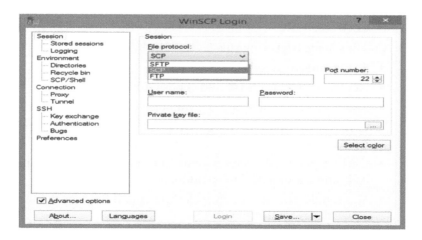

- Enter **Host name**: (192.168.10.104) IP address as we got from the previous step
- **User name:** root and **Password:** Power@1234
- This will open the WinSCP. (If it fails to check IP, user name and password once, try to log in again.)
- Now open Hands-on folder and locate to Ubidots folder and open it.
- As you can see, the second folder opens the PYTHON folder and copies UbidotsUpdateData.py file.
- Paste the copied folder/file to /root directory in WinSCP.
- To file-> Paste
- Then click on Copy, as shown below; this will copy (store) all files in the root directory.

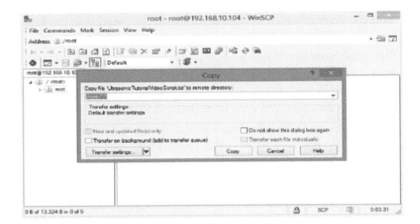

- Log in to LinkIt Smart using Putty with USB to UART converter

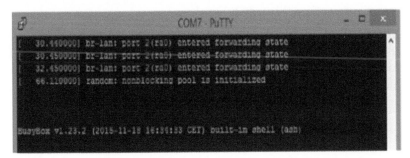

- Go to Root folder by entering command cd/root.
- Open file **ubidotsUpdateData.py** with vim editor; to do that, use the following command:

```
# Vim ubidotsUpdateData.py
```

- Replace TOKEN and VARIABLE_ID_SENSOR with your token and variable ID.
- Run the Python program using the following command; this program will receive sensor data from the Atmega32 controller and send it to the Ubidots cloud.

```
# Python ubidotsUpdateData.py
```

- Run the Sensor Sketch to read sensor data and send it to the Linux side (i.e., MT7688AN SoC)
- Open the Sketch in Arduino IDE and open program for Ultrasonic sensor.
- Connect the Ultrasonic sensor to D6 port (i.e., trigger pin goes to digital pin and echo pin goes to digital pin 7).
- Now finally go to Sketch > upload the sketch to LinkIt ONE. Then the obtained results will be visible to the users.

2.4 A Case Study on Arduino UNO Breadboard for Connecting Tiny IoT Devices

Arduino is an open source hardware for connecting tiny IoT devices to the Internet, as shown in Figure 2.4. It was developed by Massimo Banzi and David Cuartielles in 2005. Arduino coding is accessible and transferable in all languages like C++, Java, and Python.

2.4.1 Getting Started

2.4.1.1 Software Installation

- Arduino (v.1.5+) IDE (Integrated Development Environment) software
- Drivers (FTDI) (included during installation)

2.4.1.2 Hardware Components

- Arduino board (UNO R3)
- Arduino data sheet
- Analog I/O, digital I/O, serial components
- Jumper wires
- Some required peripherals (Table 2.1)

TABLE 2.1

Arduino Basic Components

Name	Type	Function	Notes
Push Button	Digital Input	Switch—Closes or opens circuit	Polarized, needs resistor
Trim Potentiometer	Analog Input	Variable resistor	Also called a Trimpot.
Photo Resistor	Analog Input	Light Dependent Resistor (LDR)	Resistance varies with light.
Relay	Digital Output	Switch driven by a small signal	Used to control larger voltages
Temp Sensor	Analog Input	Temp Dependent Resistor	Used to measure temp.
Flex Sensor	Analog Input	Variable resistor	Res. Changes with degree of bend
RGB LED	Dig & Analog Output	16,777,216 different colors	Ooh... So pretty.

2.4.2 Experimental Setup

Let us start the programming with Arduino board. Before that, make sure that all the required and supported hardware components are collected and properly connected according to the required programming environment. The following figure depicts the Arduino board with experimental setup.

2.4.3 Arduino C Specific Functions

- PinMode (pin, mode)—It describes the identified pin for input and output along with specified mode.
- DigitalWrite (pin, value)—It sends voltage level to the selected pin.
- DigitalRead (pin, value)—The designated pin may read the current voltage level.
- Analog versions of above—0 to 1023 is the range of analogRead.
- Serial commands—print, write, println.
- The assigned laptop/PC communicates to the Arduino micro controller through the serial port via a USB-serial adapter.
- After that, checkout the drivers are properly installed or not.
- If want to check the Arduino on which place the port is available. Go to My computer— Manager—Device Manager—COM port. The entire process is as shown in Figure 2.3.

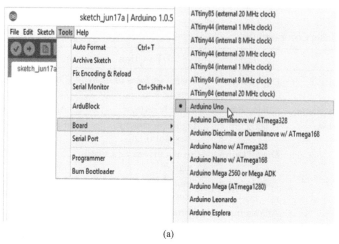

(a)

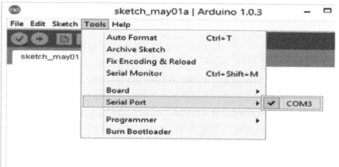

(b)

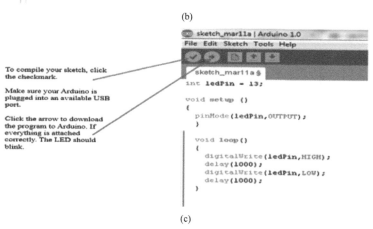

(c)

FIGURE 2.3
Choosing (a) board; (b) serial port; (c) compilation.

This is a simple C program code for Blinking LED light program.

```
int led = 13;
Void setup ()
        {
                Pin Mode (led, OUTPUT);
        }
Void loop ()
        {
                DigitalWrite (led, HIGH);
                Delay (1000);
                DigitalWrite (led, LOW);
                Delay (1000);
        }
```

The SERIES circuit along with LED and resister are wired one by one. These are the list of steps to be followed for making the circuit more acknowledged.

1. Run the 5v red wire and red strip at the place of BB on Arduino board.
2. Run the GROUND (GND) black wire and blue strip at the place of BB on Arduino board.
3. Make sure the connections of H22 and H21 of LED with the LED H22 as longer leader.
4. After that place the resister on location I21 and I11. Here row 21 is sharing both resister and LED.
5. Run the red wire at port 13 to F22 on Arduino.
6. Run the black wire to the blue strip from J11.

2.5 A Case Study on LinkIt Smart AWS IoT

In this case study shows for the IoT developers how to use the Amazon web services and Internet of Things applications in an experimental way, step by step. Thes following steps illustrate the instructions for preparing the LinkIt Smart AWS IoT.

2.5.1 LinkIt Smart AWS IoT Instructions

Step 1: Create a Things in AWS IoT and Generate and download AWS IoT device certificates.

Step 2: Create and attach a policy to the generated certificate.

Step 3: Connect LinkIt Smart to Internet.

Step 4: Transfer the AWS IoT device certificates from your computer to LinkIt Smart using Winscp.

Step 5: Run the code.

Step 6: Create a rule for the AWS IoT Thing and upload the data points from LinkIt Smart to DynamoDb.

Step 1: Create a Things in AWS IoT

1. Go to Amazon web services

   Choose AWS IoT from Internet of Things menu bar.

2. To Connect an IoT device to AWS IoT, Create a Thing; this Thing will represent your device.

   a. To create a Thing, Navigate from Dashboard to Registry.

   b. In Registry select Things and click on Create.

   c. Give any Name to the Thing. Then click Create Thing.

   d. From your Things details page navigate to Security and click Create certificate.

   e. Download device certificate and the public, private keys that are generated. Then click on Activate.

   f. Then click on Attach a policy.

3. Create and attach a policy to the device certificate.

   These are the steps to be followed for creating and attaching the policy to the certificate of chosen device.

   a. In the attach policy screen, click on Create new policy.

   b. In the Create a policy screen, Enter the Name of the Policy.

   c. In the statements fields give the action that the policy allows and the ARN of the resource for which is created. In the below example and All actions and can be performed by all resources to which this policy is attached to.

   d. Click on Allow check box and click on Create to create the policy.

   e. Navigate to your device certificate and click on the Actions drop-down and select Attach policy.

   f. Select your policy from the list and click on Attach.

Step 2: Connect LinkIt Smart to Internet

Connecting to LinkIt Smart Using Serial Port

1. **Install Drivers for USB to UART (Serial) Converter**

   a. Download The USB to UART converters driver from CH34x_Install_Windows_v3_4.zip.

   b. Extract the contents and install driver.

   c. Connect the USB to UART converter to Computer and open Device Manager (Open Control Panel, Search for Device Manager).

   d. Click on Ports (COM and LPT), and check for USB to UART, and note the COM port number.

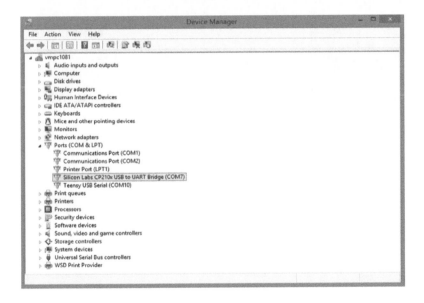

2. **Connect LinkIT Smart Board's Debug port to USB to UART converter**

   a. Connect GND of Debug to GND of USB to UART converter.

   b. Connect Tx of Debug port to Rx of USB to UART converter.

   c. Connect Rx of Debug port to Tx of USB to UART converter.

3. Download and Run Putty from https://the.earth.li/~sgtatham/ putty/latest/x86/putty.exe

4. Select Connection Type as Serial and set speed as 57600 and the change COM1 to the port number of the USB to UART converter from device manager

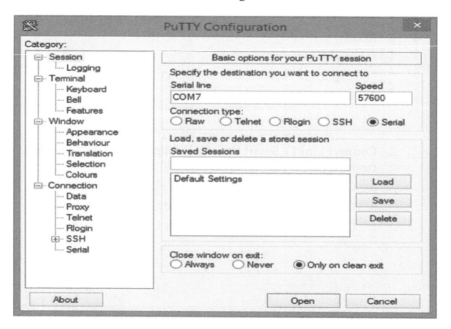

5. Click Open; the following window will appear:

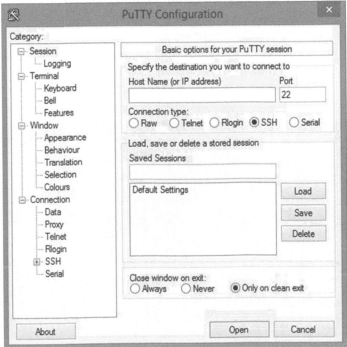

6. Connect the USB cable to LinkIT Smart Board and system. The Serial window will show the booting information; wait for 30 seconds and press Enter. The board will enter into Command Line interface.

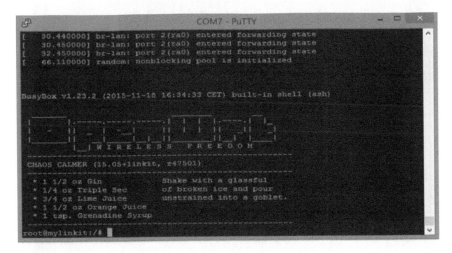

Configuring the Wi-Fi connection using the script.

1. Go to Root folder by entering command cd /root.

2. Run the Wi-Fi configuration script by running the command sh Wireless_Configuration.sh.

   Enter SSID, Password and Encryption type as shown in below manner. (If there is no password then just press Enter when password is asked and select the encryption as None.)

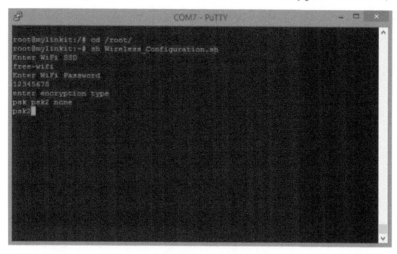

3. After running the script the board will restart; to check the connection with router run the command ifconfig, scroll up and check the IP address of apcli0 interface.

Check for the IP address assigned by the router, If the IP address is not assigned, check the SSID, password and encryption type or try changing the encryption type and run the script again.

4. To check the Internet connection run the command ping www.google.com, If there is no reply, check the Internet connectivity at the router.

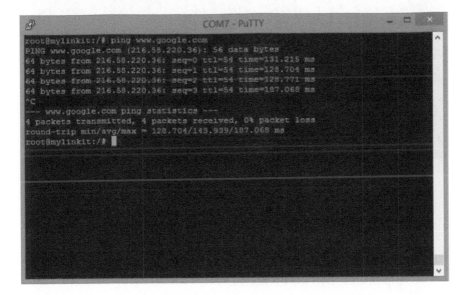

Step 3: Install AWS IoT SDK and Transfer the AWS IoT device certificates from your computer to LinkIt Smart using Winscp.

Login to LinkIT Smart thought serial USB to get the IP address of the LinkIT smart, which is assigned by the Wi-Fi router/mobile hotspot.

Use ifconfig command to get the IP address as shown below.

```
# ifconfig
```

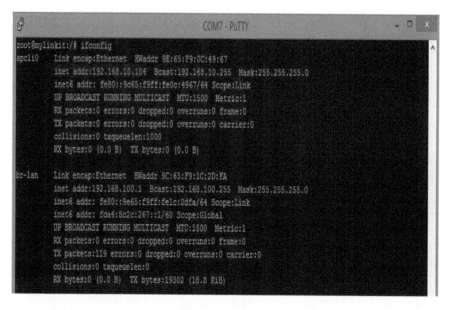

This command will give you IP address as shown in image the highlighted part (i.e. **apcli0**

Inet addr: 192.168.10.104).

Your board's IP address may be different than this IP.

Install the AWS Iot SDK using the following command:

pip install AWS IoT Python SDK

Open WinSCP and click on New session.

Select file protocol as SCP as shown in image below:

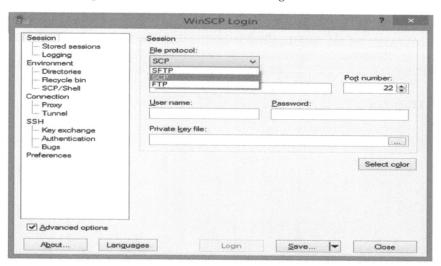

Enter **Host name**: (192.168.10.104) IP address as we got from pre-views step.

User name: rootroot

Password: admin@1234

This will open the WinSCP. (If this fails, check IP, user name and password once and again try to log in again.) Copy the contents of root Hands On in computer to root folder in LinkIt Smart DUO. Locate and navigate to the downloaded certificates on the computer.

Copy the files to the **certs** folder in the **AWS-IoT-Temperature-Humidity** or **End-to-End** folder for AWS IOT Temperature Humidity Practical or End to End Garbage collector Practical respectively.

Step 4: Run the Python code

Run the code in corresponding folder and check the thing shadow in AWS IoT.

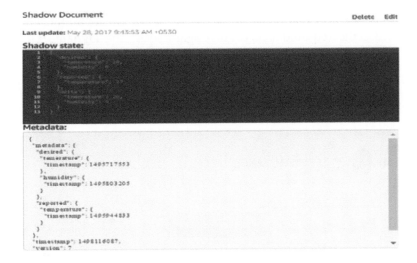

Step 5: Create a rule for the AWS IoT Thing and upload the data points from LinkIt Smart to DynamoDb.

Create a rule to take some action on the data that we get from our IoT device.

To create a rule for the AWS IoT Thing:

1. In the AWS IoT **Dashboard**, Click on **Rules** tab
2. **The AWS IoT rule contains 5 components**
   a. Name (for Rule).
   b. Overview (overview of the Rule).

     c. SQL Query

     d. SQL name of the Version.

     e. Action to be used for appropriate queries

3. **In this experiment, use Dynamo Db as our action.**

     a. Select DynamoDb from AWS.

     b. Select the **Create table**. Then DynamoDb consists of stored table.

     c. The created primary key is consisting of two or more attributes that solely identifies table contents, raw data and format the data each and every partition. Then click **Create** button, once you have enter the primary key and table name.

     d. Once table has been created, may see the properties of the created table.

4. If the table has been created, then go for getting the permissions of Amazon Web services IoT thing, need to create a **Role** along with policy. At last the created table has stored in Dynamo Db.

     a. Go to **Create** page menu, in that select **Create** button, enter the required fields. Then again select the **Dynamo Db**— actions, in that select the **Configure Actions.**

     b. Now make sure to fill Dynamo Db Configurations values. Then Go to **Role** in that select **Add Action**. Finally select the **Create Rule.**

     c. If you want to send data to be inserted in DynamoDb from the allocated device.

At last, the data has been saved by DynamoDb can execute the below SQL query.

   SELECT **state.updated.temp** FROM '**$awservice/IoTthings/myThing WHERE state.updated.temp > 24.**

2.6 A Case Study on RESTful CoAP Protocol for IoT Applications

CoAP stands for Constrained Application Protocol. CoAP is an application layer protocol like HTTP and UDP, which are mainly used for tiny IoT based Applications. CoAP mainly works on top of HTTP functionalities. It is the subset of REST Architecture. REST defines simply the method of exchange of data between clients and servers over HTTP protocol. REST has the greatest principles like cacheable mechanism, stateless, and uniform interface. Like REST architecture style, CoAP is also using URIs for identifying resources through representations. These URIs can be treated as nouns and HTTP methods like GET, PUT, UPDATE, and DELETE treated as verbs. REST is an architectural style for connecting distributed hypermedia systems. A CoAP with REST functionalities then called as RESTful CoAP protocol. In addition, other protocols for IOT applications include Message Queue Telemetry Transport (MQTT), low power wireless communication protocol (Z-wave), RACS, AMQP, and XMPP. Dissimilar to HTTP-based conventions, CoAP works over UDP; what's more, it utilizes a basic retransmission component rather than utilizing complex log control as utilized as a part of standard TCP.

Figure 2.4 shows CoAP Protocol structure format. It is a protocol applied in application layer that allows simple electronic devices to communicate over the Internet. It is designed for small devices with low-power sensors and actuators for managing things over the Internet through standard network properties. CoAP is designed for use with constrained nodes and networks. CoAP resembles with HTTP by adopting the REST model with GET, POST, PUT, DELETE methods. The following four types of messages are used for exchange of information using CoAP protocol.

1. **Confirmable (CON)** messages dependably convey a demand or reaction and require an Acknowledgment.

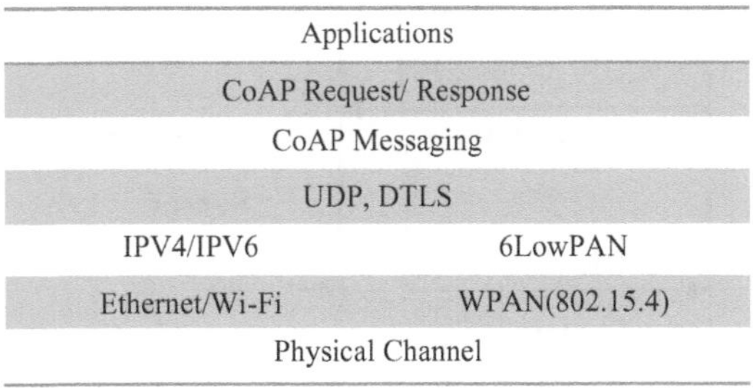

Applications	
CoAP Request/ Response	
CoAP Messaging	
UDP, DTLS	
IPV4/IPV6	6LowPAN
Ethernet/Wi-Fi	WPAN(802.15.4)
Physical Channel	

FIGURE 2.4
CoAP structure format.

2. **Non-Confirmable (NON)** messages are utilized for frequently rehashed messages and do not require an Acknowledgment (e.g., memberships of perusing a sensor).
3. **Affirmation (ACK)** messages recognize CON messages and should convey a reaction or be void.
4. **Reset (RST)** messages sent on the off chance that a CON message is not getting legitimated or some setting is absent.

CoAP structure mainly consists of various constrained devices, server, and a proxy. Through CoAP protocol, these constrained devices will communicate with each other. This structure clearly shows that for client and server interaction HTTP protocol is needed; for the constrained devices communication, CoAP is needed. Furthermore, REST style, as mentioned, depicts all communication possible between the Internet and constrained environments using REST design principles.

Figure 2.5 shows Request and Response transmission of messages between Client and Server using HTTP and CoAP protocols. A Client sends a request as HTTP Request for and a server reacts as HTTP response. This procedure is called as "Messaging." The HTTP Client sends a request for Temperature incentive to the HTTP server utilizing GET technique, and the HTTP Server

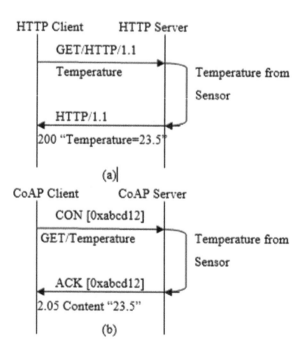

FIGURE 2.5
(a) Message Transfer utilizing HTTP; (b) Message Transfer utilizing CoAP.

TABLE 2.2

Comparison of Different Messaging Protocols

	M2M Communication	Asynchronous Messaging	Security	Resource Discovery	Message Caching
MQTT	Yes	Yes	TLS	No	Yes
AMQP	Yes	Yes	SASL, TLS	No	Yes
DDS	Yes	Yes	SRTP	Yes	Yes
XMPP	Yes	Yes	SASL, TLS	Yes	Yes
MQTT-SN	Yes	Yes	SASL, TLS	No	Yes
CoAP	Yes	Yes	DTLS	Yes	Yes

reacts to the client with the aftereffect of temperature incentive to the HTTP Client with fruitful message utilizing response code 200. The CoAP Client asks for a temperature incentive to CoAP server utilizing CON Message and GET method. The CoAP server sends the response of temperature value to the CoAP Client with the response code 2.05 and ACK Message. This process distinguishes which message transactions are done through message-id [0xabcd12], and it enables the reliable transmission.

Table 2.3 demonstrates the sensor resources that are given on the sensor hubs. For instance, a CoAP "CON+GET" request has been used to recover dampness from a sensor hub. The CoAP server keeps running on the sensor hub and sends the CoAP response with the moistness piggy-sponsored on an ACK message. Here /r resource uses GET method for getting patients' temperatures and will display in the mobile/web-based application.

2.6.1 RESTful CoAP Implementation

This section completely describes the CoAP implementation for accessing resources in the Internet of Things over cloud environment using Contiki operating system and Cooja simulator. Contrasted with Transmission Control Protocol (TCP) based conventions, User Datagram Protocol (UDP) based conventions perform better for compelled systems, since they utilize a lower number of messages while recovering the resources. Yet, contrasted

TABLE 2.3

CoAP Resources on Sensor Nodes

Resource	GET	PUT	Comments
/st	X		Temperature
/sh	X		Humidity
/sv	X		Voltage
/r	X		Voltage, Humidity, Temperature together
/l	X	X	LEDs
/ck	(X)	X	AES Encryption Key

with TCP-based HTTP convention, CoAP with UDP is a solid convention since it has straightforward retransmission capacity.

2.6.1.1 Porting Libcoap to TinyOS

The server and client components used for working CoAP protocol on TinyOS nodes candemonstrate the usage through Libcoap. At present, GET and PUT methods are used as selected resources for the server, while POST and DELETE methods are not supported to implement in this server's components. Typically, TinyOS is component-based architecture, so if any change occurs on resources amid runtime it is not conceivable and afterward will not permit creation or evacuation of resources. In IEEE, 802.13.4, the maximum number of messages requires for transfer payload, is more when compared with the CoAP application protocol. Because it does not require many numbers of resources for transfer payload and support, block-wise transfer is required for TinyOS. The Low-control IP usage for TinyOS is utilized to transmit and get the message over HTTP for to give IPv6 to CoAP on TinyOS hubs. In [12] are shown all the installation instructions for CoAP on TinyOS.

2.6.1.2 Resource Access

However wired over the Read Resource interface, the server part takes the nonexclusive segments of the read comparison just resources on the TelosB bit (i.e., Temperature/Humidity sensor). The following example demonstrates the GetResource interface:

```
interface GetResource
    {
Command error_t get (Coap_tid, id);
event void getDone (error_t result,
            Coap_tid id,
            Unit8_t asyn_message,
            Unit8_t Val,
            Unit8_t buflen);
event void getPreAck (coap_tid, id);
}
```

In order to read sensor nodes, the centralized GetResource interface has been acquainted with giving various free cases of a similar interface for every single accessible resource. It limits the code space, takes out fan-outs and transportability, and permits distinctive kinds of hubs for changing the wiring. Two events, getDone () and getPreAck (), and the Get () command are the listed commands the GetResource interface currently provides. Figure 2.6 shows the implementation of a CoAP GetResource.

   Get () command is used for GetResource while the Put () command is used for WriteResource. Both readable and writable resources like LED allow

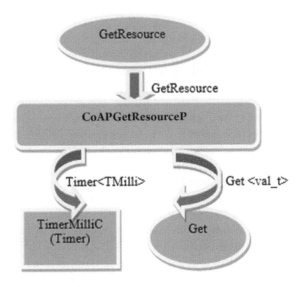

FIGURE 2.6
CoAP GetResource implementation.

using ReadResource and WriteResource interface respectively. Figure 2.7 shows that the calculations have been performed in TranslateC components using Get interface. Therefore, the recovery information can be gotten from the customer side and must be isolated with 100 to get the relating floating-point value. The temperature will get from smart home applications and it

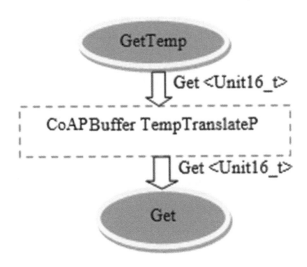

FIGURE 2.7
CoAP data translation.

reads the temperature from the room as 16 using Get <Unit16_t>. Therefore, the CoAP implementation on sensor nodes is applicable to find whether the getting resource is less energy consumable or not. Even though the resources are not less energy consumable, it will also check the reliable nature of the RESTful Resources.

Here it takes a portion of the underlying outcomes taken to assess libcoap usage on TinyOS and used one CoAP server, two TelosB nodes and one router for connected via IEEE 802.13.4 wirelessly. Table 2.4 demonstrates a couple of results while recovering diverse resources on the CoAP server to assess the execution. To quantify the recovery time, the CoAP Client on the host machine has appeared in the aggregate time taken to recover a given resource. The measure of all messages can be accomplished through an aggregate number of bytes transmitted (ACK, CON, preACK, etc.) traded amid a recovery of a resource. The executed resources '/st' and '/r' are dealt with as conceded reactions (in light of the fact that in coap-03, the RESPONSE-TIMEOUT is low) and different assets were actualized as prompt reactions. The conceded reaction message demonstrates the most noteworthy number of recovery time and bytes transmitted over a resources execution.

The resources are taken from different categories of the RESTful objects and identify the corresponding HTTP methods supported by REST architectural style. Nevertheless, retrieval time and number of bytes transmitted has been measured on corresponding resources respectively. In all of these resources, '/l' has taken the lowest retrieval time, even though the number of bytes is moderate. However, CoAP resources are considered for applying on sensor nodes to obtain retrieval time and number of bytes transmitted has measured on corresponding resources respectively. In all of these resources '/l' has taken the lowest retrieval time and '/r' taken the most retrieval time, even though the number of bytes is more, as mentioned in Table 2.4.

Figure 2.8 demonstrates the memory use in bytes while executing on a CoAP server. Using blip 1.0 the comparison of different server nodes is running in ubiquitous healthcare systems' UDPEcho. In addition, the other comparison of CoAP server nodes with different types of resources is starting from "CoAP" to "CoAP+ALL." Here "CoAP+ALL" has been mentioned for showing ROM usage running at all resources listed in Table 2.2.

TABLE 2.4

CoAP Resources on Sensor Nodes

Resource	Type	Retrieval Time	Num. of Bytes Transmitted
/st	GET	286.14 ms	221 bytes
/sh	GET	138.56 ms	116 bytes
/sv	GET	91.69 ms	115 bytes
/r	GET	356.89 ms	219 bytes
/l	GET	68.45 ms	118 bytes
/l	PUT	70.22 ms	119 bytes
/ck	PUT	100.51 ms	144 bytes

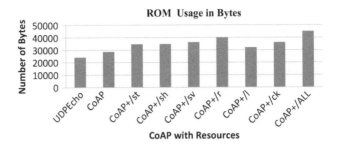

ROM Usage in Bytes

FIGURE 2.8
Memory usage (bytes).

2.6.2 CoAP vs. HTTP

This section completely depicts the CoAP versus HTTP comparison and also shows that to assess the execution of CoAP, utilizing some underlying tests taken from various resources on different sensor nodes, comparison was done between HTTP over TCP and HTTP over UDP protocols for retrieval of data. In Table 2.5, sno 1–5 shows HTTP on TCP-based protocols and sno 6–7 shows HTTP on UDP-based protocols. These underlying tests were completed by utilizing two workstations arranged as Client and Server. With GPRS organization these two workstations are associated. Therefore, we need to execute HTTP with CoAP over TCP and UDP protocols, it may require an obliged external connection of multi-trust WSN and RTT of GPRS within the range of 800 ms. These two workstations have arranged to run on HTTP construct conventions as well as with respect to CoAP-based conventions. To get the estimation of temperature from PC, both HTTP GET and CoAP asks for are utilized.

Contrasted with TCP-based conventions, the UDP-based conventions can retrieve data using only two messages. The following are the resulted metrics that should consider the evolution of CoAP protocols for energy efficiency applied to different types of resources to get the data to be retrieval.

TABLE 2.5

Response Time (Seconds)

S.no	Access Medium	Time (Seconds)
1	Apache2(Epiphany)	31.985
2	Apache2(Firefox)	38.774
3	Apache2-bareHTTP client	3.032
4	Apache2-wget	2.660
5	bareHTTP(TCP)	3.076
6	bareHTTP(UDP)	1.104
7	CoAP	1.012

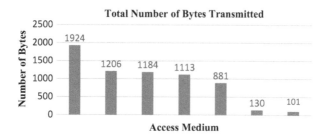

FIGURE 2.9
Total number of bytes transmitted.

1. **Response Time:** The time is taken between client request to server response or closing connection.
2. **Total number of bytes transmitted:** By a given response time see in Figure 2.9.
3. **Header Overhead:** At each layer, the number of bytes has separated.

Whenever you are using a web browser, if you want to download a file from that browser, you need to use the HTTP GET command. Figure 2.9 shows how many numbers of response time and the number of bytes transmitted over the resources. Here these resources are transmitted through different web browsers like Apache2-firefox, Epiphany and Apache2-wget and so on, and a comparison is done based on the HTTP on TCP and HTTP on UDP protocols. Finally, the authors observed that UDP-based approaches taken less number of packets, access time and number of bytes transmitted while compared with TCP-based access techniques. Table 2.6 clearly shows that TCP-based HTTP has taken the maximum number bytes from each layer while UDP has taken a lesser number of bytes because it takes low payload data and header size. This comparison has done at each layer to transmit two HTTP messages, and it was observed that UDP/CoAP takes fewer bytes compared with other protocols, as shown in Table 2.6.

TABLE 2.6

Division of Bytes in Each Layer

Header	TCP/HTTP	UDP/HTTP	UDP/CoAP
Data	4	4	2
CoAP/HTTP	181	41	17
UDP/TCP	340	16	16
IP	200	40	40
Link layer	160	32	32

These results show that compared to HTTP on TCP-based protocols, the CoAP on UDP protocols performs well for constrained devices whenever retrieving resources. The reason is that it occupies fewer messages. UDP is a reliable protocol because it has its own retransmission mechanism compared with TCP.

2.7 Conclusion and Future Work

The Internet of Things (IoT) mainly acts as a communication medium between things and the Internet. There are some estimates that there will be 50 billion connected devices to the Internet by 2020. The Internet of Things based use cases, case studies and examples are applied on the sectors of healthcare, industry 4.0, logistics 4.0, smart city, smart energy, smart grid, smart building, smart transport management, smart retail, and more. These all are fully operational, customer manageable, business supportable, and scalable. Hence there is a gap between the IoT applications side from researchers' side. In this chapter, we totally focused on various supporting programming technologies for IoT applications in an exploratory case study way. In total, six case studies have been studied for to fill the gap between the researchers side to IoT platforms side.

The first case study of this chapter was on ThingSpeak, an IoT cloud platform for IoT analytics. This is the medium to collect and analyze the data, and to visualize the collected data within a MATLAB environment. This case study demonstrates how to post the collected data from external sources to a ThingSpeak channel. For that use the ESP8266 board followed by the ThingSpeak-IoT cloud platform for both posting data as well as retrieving data from a created channel.

The second case study fully focused on sending sensor data to the Ubidots cloud platform using Wi-Fi through RESTful API. Here, the sensor data coming from various IoT devices like sensors, actuators, global positioning systems (GPSs), laser scanners and RFID cards was sent to Ubidots, a cloud platform for analysis and visualizations at online.

The third case study was on Arduino. Arduino is an open source hardware for connecting tiny IoT devices to the Internet. It was developed by Massimo Banzi and David Cuartielles in 2005. Coding is accessible and transferable in all languages like C++, Java, and Python.

The fourth case study dealt with LinkIt Smart AWS IoT. This case study showed IoT developers how to use Amazon web services and Internet of Things applications in an experimental way. The LinkIt Smart AWS IoT platform was shown to the developers in a step-by-step manner.

The last case study studied the RESTful CoAP application layer protocol for tiny IoT-based applications. CoAP protocol consumes less energy for getting IoT resources.

So far, we've discussed how the IoT programming technologies have been applied in cloud platforms and IoT hardware devices from an experimental point of view, with case studies. The following are the emerging areas that may be considered for future work:

1. The future of the Internet of Things is closely related to the future of artificial intelligence.

   In future, by applying any of the Machine Learning (ML) algorithms in IoT platforms and performing intelligent data analysis. The ML algorithms like Support Vector Machines (SVM), Naive Bayes, Linear Regression, Decision Trees, Neural Networks, K-Means, Gaussian Mixture, and Hidden Markov Model are popular for performing intelligent data prediction and analysis.

2. Data security in IoT is the most emerging topic in future to do research.

3. Handling unstructured data is also a big challenge, so there is scope to consider Big Data Analytics (BDA) in IoT.

4. Block chain techniques trends is the future of IoT.

The authors conclude that, after brief study of these case studies, this chapter may fill the gap between the users and IoT-based application developers in a programming perceptive way.

References

1. D. Singh, G. Tripathi, and A. J. Jara "A survey of Internet-of-Things: Future vision, architecture, challenges and services." In. *Internet of Things (WF-IoT), 2014 IEEE world forum on. IEEE*. 2014. pp. 287–292.
2. S. López, D. C. Ranasinghe and M. H. Duncan McFarlane, "Adding sense to the Internet of Things an architecture framework for smart object systems," *Personal Ubiquitous Computing*, 2012, vol. 16, pp. 291–308.
3. S. Balakrishna and M. Thirumaran "A RESTful CoAP protocol for Internet of Things" In. *Proceedings of 7th International Conference on Informatics Computing in Engineering Systems (ICICES)*, IEEE, 2018. pp. 1–6.
4. J. Gubbi, R. Buyya, S. Marusic, M. Palaniswami, "Internet of Things (IoT): A vision, architectural elements, and future directions," *Future Generation Computer Systems*, 2013, 29(7) pp. 1645–1660.

5. S. Takatori, S. Matsumoto, S. Saiki, M. Nakamura, A proposal of cloud-based home network system for multi-vendor services In. *IEEE/ACIS International Conference on Software Engineering, Artificial Intelligence, Networking and Paral-lel/Distributed Computing (SNPD)*, IEEE, 2014, pp. 1–6.

6. T. Kim, C. Ramos, S. Mohammed, Smart City and IoT, *In Future Generation Computer Systems*, 2017, Vol. 76, pp. 159–162.

7. D. Bandyopadhyay and J. Sen. "Internet of things: Applications and challenges in technology and standardization." *Wireless Personal Communications*, 2011, Vol. 58, no. 1, pp. 49–69.

8. L. Atzori, A. Iera, and G. Morabito. "The internet of things: A survey." *Computer Networks*, 2010, Vol. 54, no. 15, pp. 2787–2805.

9. S. Mijovic, E. Shehu, and C. Buratti "Comparing application layer protocols for the Internet of Things via experimentation." *Research and Technologies for Society and Industry Leveraging a Better Tomorrow (RTSI)*, 1–5, 2016.

10. Z. B. Babovic, J. Protic, and V. Milutinovic, "Web performance evaluation for Internet of Things applications" *IEEE Access*, 2016, vol. 4, pp. 6974–6992.

11. S. Cirani, M. Picone, and L. Veltri. "CoSIP a constrained session initiation protocol for the Internet of things" In *European Conference on Service-Oriented and Cloud Computing*, 2013. 13–24. Springer.

12. V. Kadam, S. Tamane, V. Solanki, "Smart and connected cities through technologies," In *Big Data Analytics for Smart and Connected Cities*, 2019. 1–24. IGI Global, Hershey, PA.

13. D. D. Sanju, A. Subramani and V. K. Solanki "Smart city: IoT based prototype for parking monitoring & parking management system commanded by mobile app" In *Second International Conference on Research in Intelligent and Computing in Engineering*, 2017.

14. R. Dhall & V. K. Solanki, "An IoT based predictive connected car maintenance approach" *International Journal of Interactive Multimedia and Artificial Intelligence* 2017, Vol. 4. 3 (ISSN 1989-1660).

15. V. K. Solanki, M. Venkatesan & S. Katiyar "Conceptual model for smart cities for irrigation and highway lamps using IoT," *International Journal of Interactive Multimedia and Artificial Intelligence*, (ISSN 1989-1660), 2018.

16. V. K. Solanki, M. Venkatesan & S. Katiyar "Think home: A smart home as digital ecosystem in circuits and systems," *Scientific Research Publishing Inc.* 2018, Vol. 10, No 07, ISSN 2153-1293.

17. S. Balakrishna and M. Thirumaran "Semantic interoperable traffic management framework for IoT smart city applications," *EAI Endorsed Transactions on Internet of Things*, 2018, Vol 4, Issue 13, pp. 1–17. doi:10.4108/eai.11-9-2018.15548

18. V. K. Solanki, S. Katiyar, V. Bhaskar Semwal, P. Dewan, M. Venkatesan, N. Dey "Advance automated module for smart and secure city" in *ICISP-15*, organised by G.H. Raisoni College of Engineering & Information Technology, Nagpur, on 11–12 December 2015, published by Procedia Computer Science, Amsterdam, the Netherlands: Elsevier, ISSN 1877-0509

19. R. T. Fielding, 2000. *"Architectural Styles and the Design of Network-Based Software Architectures,"* doctoral dissertation, Irvine, CA: University of California.

3

Design and Construction of a Light-Detecting and Obstacle-Sensing Robot for IoT—Preliminary Feasibility Study

Mohammad Farhan Ferdous, Prayag Tiwari, and V. B. Surya Prasath

CONTENTS

3.1 Introduction

Automation is pervasive in the current decade, and nowadays every system is automated in order to face new challenges. The advantages of such automated systems include reduced manual operations, more flexibility, reliability, and guarantee of higher accuracy. Due to this trend, the majority of the modern fields prefer automated control systems; especially in the field of electronics, automated systems are proven to provide good performances. Further, unmanned systems play very important role to minimize of the risk of human life; for example, drones/unmanned aerial vehicles (UAVs) are increasingly utilized in modern wars. Recent advancements in the field of robotics have increased the demand for automated systems that can self-sustain, detect, and manoeuver tasks efficiently.

Automation technology in off-road equipment is an active area of research and several technological hurdles are yet to be solved [1]. One of the biggest hurdles is reliable detection of obstacles and recognition. There have been efforts to model obstacle detection using visual servoing. Using a novel autonomous control architecture, specialized sensing, combined with manipulation and visual servoing as well as Bayesian classification, the nomad robot found and classified five indigenous meteorites during an expedition to the remote site of Elephant Moraine in 2000 [2]. The primary reason for utilizing such a mobile robot is the capability of it to execute one or more tasks repeatedly with high speed as well as accuracy [3]. In [4] a design for the controller was undertaken with stable walk with defined dynamic and static parameters. Further, the authors of [4] humanoid locomotion as a hybrid system proposed the canonical equation for the universal for momentum of different join angles to produce exact human locomotion. A key capability required by an autonomous mobile robot is the possibility to navigate through an area. The one of the goals of robot navigation is obstacle avoidance. The mobile robot must travel from a start point to a target point while being able to detect and avoid the obstacles [5]. Hence the robot always keeps a safe distance from the obstacles in order to travel from the start point to the goal point. The navigation of the mobile robot can be more efficient if the robot can detect the distance of the obstacle from the robot. Thus, good map constructions of the atmosphere and using a proper sensor that gives accurate data are essential to repel from obstacle collisions. The robot should able to process the data from the sensor as quickly as possible when it detects an obstacle [5], so that it can escape the obstacle and go to a new safe path. Both the robot and the obstacles may be damaged due to obstacle collision. There are considerable interests to build a mobile robot to use in space, military, office, hospital, and industry [6]. It also can be used to imitate human jobs like cleaner, postman, waiter, security guard, and more.

The field of robotics encompasses a broad spectrum of technologies in which computational intelligence is embedded in physical machines, creating systems with capabilities far exceeding the core components alone [7]. In a smart and connected society driven by Internet of Things (IoT) devices [8–10], robotic systems will play important roles in enabling smart cities [11–13]. Such robotic systems are then able to carry out tasks that are unachievable by conventional machines, or even by humans working with conventional tools. The ability of a machine to move by itself—that is, "autonomously"—is one such capability that opens up an enormous range of applications that are uniquely suited to robotic systems.

This research is a recent advancement in automation, healthcare, and space exploration [14]. In the field of mechatronics systems, the worth and value of automated systems is very high and gives good performance. In a war situation, these robots are very useful to reduce the human losses and money. Example of this system is target detecting. The automated system is here for detecting the targets. It is developed using an embedded system with a microcontroller. The mobile robot vehicle moves whenever required. The push recovery is still challenging, because the robot does not have as much physical strength and decision-making capability that normal humans have [15]. To be useful in the real world, robots need to move safely in unstructured environments and achieve their goals despite unexpected changes in their surroundings. The environments of real robots are rarely predicable or perfectly known, so it does make sense to make precise plans before moving.

In this work, we present a light-sensing and obstacle-detecting robotic vehicle. One of the major advantages of our automated robot is that it senses the light and directs itself toward the high-luminance areas. For this purpose, we custom-designed our robot with sensors that detect the luminance density effectively under varying illumination conditions. Further, we deployed ultrasonic sensors that allowed the robot to avoid any potential obstacles that could prevent it from reaching the high-luminance areas successfully. In this paper we describe the details of our design and construction of a light-detecting and obstacle-sensing (LiDOS) robot developed by us. Our design includes a mobile robotic vehicle where the sensors are mounted and controlled in an automatic way. From the standby mode the mobile robot vehicle is moved, and whenever an obstacle and light are detected by the photo sensor and ultrasonic sensor respectively, the information is passed to the microcontroller within the robot vehicle. Obstacle avoidance is accomplished through a combination of global and local avoidance subsystems that deal with both known and unknown obstacles in the operating area of the vehicle [16]. Based on these, the microcontroller changes the direction of the robot by driving the motors in the direction of the target. Pushes were induced from behind with closed eyes to observe the motor action as well as with open eyes to observe learning-based reactive behaviors [17]. The following components were used in the hardware side: microcontroller, battery, photo sensor or photo detectors, ultrasonic sensors, motors, and motor drivers. The hardware in this robot is a microcontroller, battery, sensors, light-dependent resistor, and motor and stepper motor drivers' body shell. This system can be used to get accurate results and to reduce human efforts.

The rest of this chapter is organized as follows: Section 2 reviews related light- and obstacle-detecting robotic vehicles systems from the literature. Section 3 provides the details of our proposed system with sensors and microcontroller. Finally, Section 4 concludes the paper.

3.2 Literature Review

In what follows, we review works related to the current study of the ultrasonic sensor and photosensor-detecting light and obstacle robotic vehicle system technology. We reviewed different obstacle-detecting robot mechanisms that have been built. These robots can be divided into two categories: either mobile or fixed robots. The purpose of this chapter is to briefly describe mobile robots that are used in difficult or dangerous environments.

3.2.1 Review of Mobile Robots

- **SR04 Mobile Robot**

  Author: David P. Anderson, Department of Geological Sciences, Southern Methodist University

  Description: The **SR04** is a small mobile robot suitable for exploring human habitats unattended. It is controlled by a Motorola HC6811 microprocessor running in an M.I.T. 6.270 CPU card, similar to the commercially available "Handy Board." Two 12-volt DC gear-head motors manoeuver the robot in a dual-differential drive configuration, balanced by a non-driven tail wheel caster and powered by a 12 volt 2.2 amp-hour sealed lead acid battery. Sensory input is provided by (in order of priority) front bumper switches, IR collision avoidance, stereo sonar ranging, photodetectors, passive IR motion detection, and shaft-encoder odometry.

- **Machina Speculatrix**

  Author: W. Grey Walter, Smithsonian Institution

  Description: These vehicles have a light sensor, touch sensor, propulsion motor, steering motor, and a two vacuum tube analogy computers. Even with this simple design, Grey demonstrated that his robots exhibited complex behaviors. He called his creation Machina Speculatrix, after their speculative tendency to explore their environment. The Adam and Eve of his robots were named, respectively, Elmer and Elsie (Electro Mechanical Robots, Light Sensitive).

  The light sensor is rigidly fixed to the steering assembly so that it is always pointing in the direction of travel. This is my third-generation design, and I have reduced the construction to the bare necessities to make it easier to build.

  Grey's turtles has a steering assembly that rotated 360 degrees and in only one direction. This requires slip rings to carry electricity to the drive motor and the light sensor. Since there is no LEGO part like this, I added a switch to detect when the steering assembly is pointed backward. This condition is used to reverse the steering motor direction, creating a windshield wiper action. The post that supports the light sensor also pushes the switch. The small wheel on the front is used for debugging and to set the initial direction.

- **Photobot**

  Authors: Dan and Caleb DeGard, Seattle, Washington

  Description: The robot was used for a robotics demonstration at a community college. Components common in the amateur robotics industry were used, such as a Parallax Basic Stamp microcontroller, servomotors and wheels from a radio control (R/C) model, a servo control chip, and a cadmium-sulfide photoresistor and potentiometer. The robot was assembled using duct tape, Velcro, and rubber bands. No soldering was required. Electrical connections were made using the wire-wrapping method. A full listing of the control program is included.

  This mobile robot was developed to serve as a demonstration at a meeting of the technology club at the Seattle Central Community College. The club had a desire to become involved in robotics, and was making their first step by inviting us to introduce them to the industry (sport, vocation, avocation, all-consuming compulsion)—initial discussions within the club were tending toward mind-numbing complexity.

  The robot consists of a Parallax Basic Stamp 2 on a carrier board, two hacked servos with wheels, a battery pack for the servos, a 9V battery for the stamp, a CdS cell, and a 10K potentiometer. In addition, a Ferrettronics FT639 chip is used to provide the square-wave signal for the servos. This chip is optional, as the Stamp can be programmed to provide the square-wave signal. It does, however, make servo operation easier, and we now tend to use it whenever we use servos with a microcontroller. There are also several LEDs mounted on the carrier board, used to track the active subroutine. These are not necessary for the operation of the robot, but they do look cool. A length of 18AWG wire is bent to form a tailskid, and a couple of rubber bands hold the Stamp carrier board on top of the servo battery pack.

- **ActivMedia Pioneer 3-DX Mobile Robot**

  The ActivMedia Pioneer 3-DX robot is a two-wheel drive intelligent mobile platform equipped with advanced devices for sensing and navigation in a real-world environment, including wheel encoders and ultrasonic sensors. The circular basis has a diameter of 50 cm and can carry up to 23 Kg. It is mainly conceived for research experiments. The robot operates as the server in a client-server environment: Its controller handles the low-level details of mobile robotics, including maintaining the platform's drive speed and heading over uneven terrain, acquiring sensor readings such as the sonar, and managing attached accessories like the Gripper. To complete the client-server architecture, the robot requires a client connection: software running on a computer connected with the robot's controller via a serial link and that provides the high-level, intelligent robot controls, including obstacle avoidance, path planning, features recognition, localization, and so on.

3.2.2 Hardware Review—Photo Light Sensors

- **Fundamental light sensors or photodetectors:** The photosensor or LDR is used to sense changes in the quantity of light in its atmosphere. It is a cadmium-supplied device and has nonlinear characteristics. However, its sensitivity to the light is quite slow, due to its response time to the light is slower compared with a photodiode or phototransistor. The value of its resistance decreases when the intensity of the light falling on the sensor increases. The behavior of the photocell can be seen as that of a light-control potentiometer.

  The photocell is interfaced with the Handy Board using a voltage divider circuit. Using this circuit, input voltage Vin is the voltage drop across a 47Kohm pull-up resistor and photosensor, and the output voltage is the voltage across the Rphot (of output voltage Sent apped between the resistors).

- **Ultrasonic distance sensor:** Parallax's PING ultrasonic sensor provides a very low-cost and easy method of distance measurement. This sensor is perfect for any number of applications that require you to perform measurements between moving or stationary objects. Naturally, robotics applications are very popular, but one can also find this product to be useful in security systems or as an infrared replacement if so desired. You will definitely appreciate the activity-status LED and the economic use of just one I/O pin.

  The PING sensor measures distance using sonar; an ultrasonic (well above human hearing) pulse is transmitted from the unit and distance-to-target is determined by measuring the time required for the echo return. Output from the PING sensor is a variable-width pulse that corresponds to the distance to the target.

  Interfacing to the BASIC Stamp and Javelin Stamp microcontrollers is a snap: a single (shared) I/O pin is used to trigger the PING sensor and "listen" for the echo return pulse. And the intelligent trigger hold-off allows the PING to work with the BS1! An onboard three-pin header allows the PING to be plugged into a solderless breadboard (on a Boe-Bot, for example) and to be connected to its host through a standard three-pin servo extension cable.

- **Ultrasonic movement detector:** A matched pair of ultrasonic transducers operating at 40 KHz will reliably detect movement 4 m to 7 m away. It uses two 9V–12V circuit boards. The circuit is crystal-locked by 40 KHz crystal for maximum efficiency. This crystal-locked ultra-sonic movement detector kit is built around a matched pair of ceramic transducers that convert movement energy to electrical energy and vice versa. Any movement in the area scanned by the transducer will be detected and a pulse produced. In this kit the pulse turns on an LED. Pads are provided to take this pulse to add-on circuits, where it may be used to turn on buzzers, lights, and so on. A PCB (printed cir-cuit board) mounted switch can be used to switch between an auto-matic reset about 0.3 seconds after the detector has been triggered or to stay latched on. The unit will reliably work from four meters to more than 8 meters, depending on the sensitivity setting and the direction of the movement.

The design was not built at once. We studied different possibilities and noticed how the thing worked and then reversed it and saw how it worked and noted down the changes. Finally we selected the optimum procedure and used the components that were best fit for the project in expenses and quality. We had studied all the relevant light sensors we needed to use in this project; finally we selected the LDR as light sensor. Then we went through the selection procedure of the range sensor for obstacle detection, and we chose the ultrasonic sensor for the obstacle detection. Finally we selected the microcontroller PIC16F676 for the project.

3.3 Design of a Light-Sensing and Obstacle-Detecting Robotic Vehicle

3.3.1 Block Diagram of the System

The block diagram in Figure 3.1 shows the sequence of the communication path of the signal generated from the photosensor (LDR) and ultrasonic distance sensor; when it comes to the pin of the microcontroller, it gives the off signal, and the motor goes in off mode and gradually stops the vehicle car.

3.3.2 Block Diagram of the Control Circuit

The block diagram in Figure 3.2 shows that whenever any obstacle is encountered in front of the range sensor (ultrasonic distance sensor) of the specimen, the sensor will give the command to the microcontroller, and when any light falls on the sensor of the specimen, it gives a command to the microcontroller, and the program in the microcontroller executes the command from the sensors and changes the output (motor).

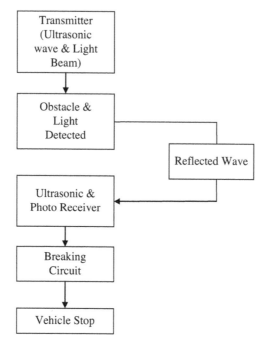

FIGURE 3.1
Block diagram of the system.

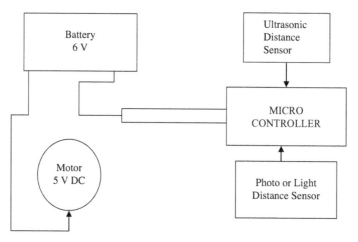

FIGURE 3.2
Block diagram of the control circuit.

3.3.3 Circuit Diagram of the Controller

Figure 3.3 shows the circuit diagram of the controller.

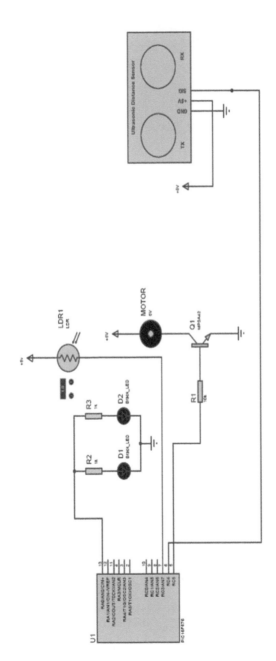

FIGURE 3.3
Circuit diagram of the controller.

3.3.4 Program

```
// define io
#define _IO_SIGNAL RA5_bit
#define _D_SIGNAL TRISA5_bit
#define _O_MOTOR RA0_bit
#define _D_MOTOR TRISA0_bit
#define _O_LED RC1_bit
#define _D_LED TRISC1_bit
#define _Solar_H RC2_bit
#define _Solar_H_Dir TRISC2_bit

void main()
{
        unsigned int EchoTime = 0, Distance = 0;

        PORTA = 0;

        PORTC = 0;

        ANSEL = 0;

        // set IO
        _IO_SIGNAL = 0;
        _O_MOTOR = 0;
        _O_LED = 0;
        _Solar_H_Dir = 1;

        // init io
        _D_SIGNAL = 0;
        _D_MOTOR = 0;
        _D_LED = 0;
        _Solar_H = 1;
        // wait time for initialize Sonar Module
        Delay_ms(500);

        while(1)
        {
                // set signal to Lo
                _IO_SIGNAL = 0;
                _D_SIGNAL = 0;
// send start signal for Tout time
                _IO_SIGNAL = 1;
```

```
                Delay_us(5);
                _IO_SIGNAL = 0;
                _D_SIGNAL = 0;
                // clear counter
                EchoTime = 0;
                Distance = 0;

                // wait for Thold time
                Delay_us(200);
                _D_SIGNAL = 1;

                // count echo time, every count equal 3us
                while(_IO_SIGNAL == 1){EchoTime++;}

                if(EchoTime > 500 && _Solar_H == 0)
                {
                        _O_LED = 0;
                        _O_MOTOR = 1;
                }
                else

                {
                        _O_LED = 1;
                        _O_MOTOR = 0;
                }

                // give delay before next measurement
                Delay_ms(100);
        }
}
```

3.3.5 Design of a Light-Sensing and Obstacle-Detecting Robotic Vehicle

3.3.5.1 *Building of Experimental Setup*

In Figure 3.4, we show our four-wheel small robotic vehicle. It is a single DC motor drive four-wheel automotive mobile vehicle. An ultrasonic distance meter sensor and light sensor are placed on the roof of the vehicle. When the toy moves in a forward direction, it always observes any obstacle and light in front and on top of it, and when detects anything or any toy car, it gives the signal to the microcontroller and stops running the motor, which is why it is called a light-detecting vehicle.

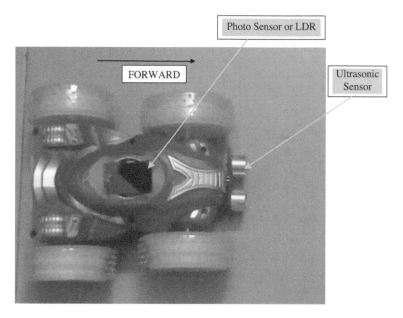

FIGURE 3.4
Light-sensing and obstacle-detecting robotic vehicle.

3.3.6 Instrument and Measurement

3.3.6.1 List of Required Components

Component	Number
Ultrasonic distance sensor	1 pcs
Photosensor or light sensor	1 pcs
Microcontroller (PIC16F676)	1 pcs
Motor (DC)	1 pcs
Power source (6VDC battery)	1 pc
Transistor	1 pcs
Diode	2 pcs
Resistor	3 pcs
Connecting wires	

3.3.6.1.1 Features

- The PING))) has only has three connections, which include Vdd, Vss, and one I/O pin.
- The three-pin header makes it easy to connect using a servo extension cable; no soldering required.

- Provides precise, non-contact distance measurements within a 2 cm to 3 m range.
- Simple pulse in/pulse out communication.
- Burst indicator LED shows measurement in progress.
- 20 mA power consumption.
- Narrow acceptance angle.

3.3.6.1.2 *Key Specifications*

- Power requirements: +5 VDC
- Communication: Positive TTL pulse
- Dimensions: 0.81 × 1.8 × 0.6 in (22 × 46 × 16 mm)
- Operating temp range: +32°F to +158°F (0°C to +70°C)

An ultrasonic sensor typically utilizes a transducer that produces an electrical output in response to received ultrasonic energy. The normal frequency range for human hearing is roughly 20–20,000 hertz. Ultrasonic sound waves are sound waves that are above the range of human hearing, and thus have a frequency above 20,000 hertz. Any frequency above 20,000 hertz may be considered ultrasonic. Most industrial processes, including almost all sources of friction, create some ultrasonic noise. The ultrasonic transducer produces ultrasonic signals. These signals are propagated through a sensing medium, and the same transducer can be used to detect returning signals. The energy bursts travel from the ultrasonic sensor, bounce off objects, and are returned toward the sensor as echoes. Transducers are devices that convert electrical energy to mechanical energy, or vice versa. Conversely, when a voltage is applied across certain surfaces of a solid that exhibits the piezoelectric effect, the solid undergoes a mechanical distortion. Transducers are also used in earphones and ultrasonic transmitters that produce a mechanical output from an electrical input. Ultrasonic transducers operate to radiate ultrasonic waves through a medium such as air. Transducers generally create ultrasonic vibrations through the use of piezoelectric materials such as certain forms of crystal or ceramic.

3.3.6.1.3 *Ultrasonic Control Basics*

Ultrasonic signals are like audible sound waves, except the frequencies are much higher. Our ultrasonic transducers have piezoelectric crystals that resonate to a desired frequency and convert electric energy into acoustic energy and vice versa. The illustration shows how sound waves, transmitted in the shape of a cone, are reflected from a target back to the transducer. An output signal is produced to perform some kind of indicating or control function. A minimum distance from the sensor is required to provide a time delay so that the "echoes" can be interpreted. Variables that can affect the operation of ultrasonic sensing include target surface angle, reflective surface roughness, or changes in temperature or humidity. The targets can have any kind of reflective form—even round objects.

3.3.6.1.4 *The Advantages of Ultrasonic Sensor*

Ultrasonic sensors have a lot of advantages for using in real application. The advantages of ultrasonic sensor are as follows:

- Discrete distances to moving objects can be detected and measured.
- Less affected by target materials and surfaces, and not affected by color.
- Resistance to external disturbances such as vibration, infrared radiation, ambient noise, and EMI radiation.
- Measures and detects distances to moving objects.
- Impervious to target materials, surface, and color.
- Solid-state units have virtually unlimited, maintenance-free lifespan.
- Detects small objects over long operating distance.
- Ultrasonic sensors are not affected by dust, dirt, or high-moisture environments.
- Ultrasonic level sensors are usually non-contact type; that is, they do not make any contact with the process fluid under level detection.
- Besides, they consist of fixed components only, hence require less maintenance.

3.3.6.1.5 *The Limitations of Ultrasonic Sensor*

Some disadvantages of ultrasonic sensor are as follows:

Ultrasonic level measurement technique cannot be suitably applied in all fields, since use of ultrasonic level sensors includes a few setbacks too. Many factors exist that have the tendency to influence the returned echo signal back to the sensor. Some of them include the following:

- Materials like powders, liquids and so on
- Heavy vapors
- Surface turmoil
- Foam
- Ambient noise and temperature

3.3.6.1.6 LIGHT or Photo Distance Sensor

This is capable of measuring the distance to objects at ranges from 10 to 80 cm. It provides a voltage output that can be connected to an analogy input on the Handy board.

The distance sensor transmits a short burst of infrared light at intervals of about 1 ms. The light source is behind one of the lenses, which focuses the light into a narrow beam.

Light reflected from an object in this beam will return to the sensor, at an angle which depends on the distance to the object. This light passes through the other lens on the distance sensor, and is focused onto an array of light detectors. Light arriving at different angles will fall on a different respective set of light detectors.

3.3.6.1.7 Devantech SRF04 Range Finder Sensor

The SRF04 was designed to be just as easy to use as the Polaroid sonar, requiring a short trigger pulse and providing an echo pulse. The SRF04 timing diagram is shown below. You only need to supply a short 10uS pulse to the trigger input to start the ranging. The SRF04 will send out an eight-cycle burst of ultrasound at 40 kHz and raise its echo line high. The minimum and maximum range describes the sensor limits at 0 degrees (straight on from the sensor), while the angle describes the rough shape of the sensor cone at one half the sensor's range. In actuality, the sensors do not detect in a perfect cone.

The echoes column lists the number of echoes recorded by the sensor. This always refers to the number of echoes recorded by the most recent ranging; with each new ranging, old values are overwritten. The range time column refers to the time to perform a ranging. The sensors using digital communication respond as soon as an echo is received.

3.3.6.1.8 Light-Dependent Resistor (LDR)

We used an LDR sensor to detect the luminance difference around the robot. These three sensors were separated by 120 degrees to easily detect the light around the robot. To isolate the LDR voltage variation from the PIC controller we used and LM229 comparator. Comparators constantly compare pairs of voltages and provide a digital indication ("1" or "0") of which voltage is higher. Using the dedicated chip frees the microcontroller, which is now interrupted only when the digital signal changes.

The LM239 is a quad, single-supply comparator. A quad can compare four different pairs of voltages.

This comparator can operate up to 36 volts (or +18 V to −18 V). Since we intend to connect it to a microcontroller, so we made this as +5 V and GND. The LM239 is pin compatible with MC3303, LM339, and LM2901 chips. Although their operating temperature ranges differ (and there are a few other differences), anyone can work fine in this project.

3.3.6.1.9 *Light or Photosensor Control Basics*

A *photoconductive* light sensor does not produce electricity but simply changes its physical properties when subjected to light energy. The most common type of photoconductive device is the photoresistor, which changes its electrical resistance In response to changes in the light intensity. Photoresistors are semiconductor devices that use light energy to control the flow of electrons, and hence the current flowing through them. The commonly used photoconductive cell is called the *Light-Dependent Resistor Or LDR.*

The most commonly used photoresistive light sensor is the ORP12 Cadmium Sulphide photoconductive cell. This light-dependent resistor has a spectral response of about 610nm in the yellow to orange region of light. The resistance of the cell when unilluminated (dark resistance) is very high at about 10MΩs, which falls to about 100Ωs when fully illuminated (lit resistance).

To increase the dark resistance and therefore reduce the dark current, the resistive path forms a zigzag pattern across the ceramic substrate. The CdS photocell is a very low0cost device often used in auto dimming, darkness or twilight detection for turning the street lights on and off, and for photographic exposure meter types of applications.

3.3.6.1.10 *Application of Photosensor or Light Sensor*

Photocells are commonly used to detect the incandescent lamp that acts as a contest start indicator. They are also used to find the light beacons marking certain parts of the board, such as the goals. While they can be used to measure the reflectivity of the game board surface if coupled with a light source such as a red LED or an incandescent lamp, the IR reflectance sensors are usually better at this function. Photocells are sensitive to ambient lighting and usually need to be shielded. Certain parts of the game board might be marked with polarized light sources. An array of photocells with polarizing filters at different orientations could be used to detect the polarization angle of polarized light and locate those board features.

3.3.6.1.11 *Advantages of Light Sensor*

- They are useful and can help many security problems.
- They protect our valuables.
- Some are easily assessable (e.g., cash points).
- They control things properly (e.g., traffic lights).

3.3.6.1.12 *Disadvantages of Light Sensor*

- They can easily be set off and cause problems.
- They can break down.

3.3.6.2 Microcontroller

In this project we used microcontroller chip PIC16F676, which has 14 pins. This intense (200 nanosecond guidance execution) yet simple-to-program (just 35 single word directions) CMOS Flash-based 8-bit microcontroller packs Microchip's great PIC® MCU engineering into a 14-stick bundle and highlights 8 channels for the 10-bit analog-to-digital (A/D) converter, 1 comparator, and 128 bytes of EEPROM information memory. This gadget is effortlessly adjusted for car, modern machines, and purchaser section level item applications that require field reprogrammability.

The features of the microcontroller are as follows:

- 128 bytes of EEPROM data memory
- Programmable draw-up resistors
- Independently selectable simple channels
- ICD2 programming support or investigating support with discretionary header connector
- Eight oscillator choices, including the Exactness 4 MHz R/C oscillator with programmable adjustment and power-on reset

3.3.6.2.1 Fundamental of PIC

PIC is a family of Harvard architecture microcontrollers made by microchip technology, derived from the PIC1650 originally developed by general instrument's microelectronics division. PICs are popular with developers and hobbyists alike due to their low cost, wide availability, large user base, extensive collection of application notes, availability of low-cost or free development tools, and serial programming (and reprogramming with flash memory) capability. PICs were also commonly used to defeat the security system of popular consumer products (pay-TV, PlayStation), which attracted the attention of crackers.

PICs have a set of register files that function as general purpose RAM; special purpose control registers for on-chip hardware resources are also mapped into the data space. The addressability of memory varies depending on device series, and all PIC devices have some banking mechanism to extend the addressing to additional memory. Later series of devices feature move instructions which can cover the whole addressable space, independent of the selected bank. In earlier devices, any register move had to be through the accumulator. To synthesize indirect addressing, a "file select register" (FSR) and "indirect register" (INDF) are used. A read or write to INDF will be to the memory pointed to by FSR. Later devices extended this concept with post- and pre-increment/decrement for greater efficiency in accessing sequentially stored data. This also allows FSR to be treated like a stack pointer.

PICs have a hardware call stack, which is used to save return addresses. The hardware stack is not software accessible on earlier devices, but this changed with the 18 Series devices. Hardware support for a general purpose parameter stack was lacking in the early series, but this greatly improved in the 18 Series, making the 18 Series architecture friendlier to high level language compilers.

3.3.6.2.2 *Instruction Set*

PICs' instructions vary in number from about 35 instructions for the low-end PICs to about 70 instructions for the high-end PICs. The instruction set includes instructions to perform a variety of operations on registers directly, the accumulator and a literal constant or the accumulator and a register, as well as for conditional execution and program branching. To load a constant, it is necessary to load it into W before it can be moved into another register. On the older cores, all register moves needed to pass through W, but this changed on the high-end cores. PIC18 cores have skip instructions that are used for conditional execution and branching. The skip instructions are "skip if bit set" and "skip if bit not set." Because cores before PIC18 had only unconditional branch instructions; conditional jumps are synthesized by a conditional skip (with the opposite condition) followed by a branch. The PIC architecture has no (or very meager) hardware support for saving processor state when servicing interrupts. The 18 Series improved this situation by implementing shadow registers, which save several important registers during an interrupt.

3.3.6.2.3 *Pin Description*

Pin Number	Description
1	Vdd—Positive Power Supply
2	RA5/T1CKI/OSC1/CLKIN—Port A
3	RA4/T1G/OSC2/AN3/CLKOUT—Port A
4	RA3/MCLR/Vpp—Port A
5	RC5—Port C
6	RC4—Port C
7	RC3/AN7—Port C
8	RC2/AN6—Port C
9	RC1/AN5—Port C
10	RC0/AN4—Port C
11	RA2/AN2/COUT/T0CKI/INT—Port A
12	RA1/AN1/CIN-/Vref/ICSPCLK—Port A
13	RA0/AN0/CIN+/ICSPDAT—Port A
14	Vss — Ground

3.3.6.2.4 *Special Microcontroller Features*

- Fail-safe clock monitor
- Power-on reset

- Power-up timer (PWRT) and oscillator start-up timer (OST)
- Two-speed start-up mode
- Programmable code protection
- Power-saving sleeping mode
- Peripheral features
  - Two 8-bit timer/counter with prescalar
  - One 16-bit timer/counter
  - High source/sink current: 25mA
  - Parallel Slave Port (PSP): 40/44 pin-device only
- Low-power features
  - Primary run (XT, RC, Oscillator, 76 μA, 1MHz, 2V)
  - RC_RUN (7 μA, 31.25kHz, 2V)
  - Timer1 oscillator (1.8 μA, 32kHz, 2V)
  - Watchdog timer (0.7 μA, 2V)
  - Two-speed oscillator start-up
- Analog features
  - 10-bit, up to 14-channel analog-to-digital converter
  - Dual analog comparators
  - Programmable low-current brown-out reset (NOR) circuitry
  - Programmable low-voltage detect (LVD)
- Oscillators
  - Three crystal modes: LP, XT, HS (up to 20 MHz)
  - Two external RC modes
  - ECIO (up to 20 MHz)
  - 8 user-selectable frequencies

3.4 Experiment Results

After building the project we were testing the robotic vehicle. Our project's first objective was design a robotic vehicle using 14 pins microcontroller (PIC16F676) that input and output pins used accurately. The working components such as ultrasonic sensor, LDR, DC motors, transistor, and other devices are connected with the microcontroller. There is a program in the microcontroller and which direction must help to detect light and obstacle of the robotic vehicle.

Our project's second objective was to develop the light- and obstacle-sensing capability of the robotic vehicle. The vehicle LDR sensor can detect to any type of light beam; to fulfill to our project objective, when a light beam falls on an LDR sensor then the vehicle will stop and the light-sensing capacity is high. We also used an ultrasonic sensor, which can sense any type of obstacle. When an obstacle is found to be in front of the robotic vehicle then it will stop travelling, else the robotic vehicle always goes forward.

The results of the work utilized the following sensors:

1. Two ultrasonic range finders mounted on the vehicle to detect obstacles and provide information to detour around the obstacle.
2. Light sources attached to a position in the room and a rotating light-detecting sensor located on the vehicle to update the absolute position.

3.5 Conclusion

The work presents a prototype design and construction of a light-detecting and obstacle-sensing robot. We showed how to build the overall circuit and programmed the PIC we chose to test the robotic vehicle. Based on the distance between the sensors and the surface of obstacles, the robotic vehicle (as the vehicle we were working on had different polarity stuff) will stop if it detects the obstacles or else seamlessly move forward. Our preliminary feasibility of building this multi-sensor based robotic vehicle shows promise, and we are currently extending this to be an Internet of Things (IoT) enabled device.

References

1. Bischoff R., "Advances in the development of the humanoid service robot HERMES," *Proceedings of the International Conference on Field and Service Robotics,* pp. 156–161, 1999.
2. Apostolopoulos D., M. Wagner, W. Whittaker, "Technology and field demonstration results in the robotic search for Antarctic meteorites," *Proceedings of the International Conference on Field and Service Robotics,* pp. 185–190, 1999.
3. Langer D., Mettenleiter, F. Hartl, and C. Frohlich, "Imaging laser radar for 3- D surveying and cad modeling of real world environments," *Proceedings of the International Conference on Field and Service Robotics,* pp. 13–18, 1999.
4. Semwal, Vijay Bhaskar, et al. "Biologically-inspired push recovery capable bipedal locomotion modeling through hybrid automata." *Robotics and Autonomous Systems* 70 (2015): 181–190.
5. Corke P., G. Winstanley, J. Roberts, E. Duff, and P. Sikka, "Robotics for the mining industry: Opportunities and current research," *Proceedings of the International Conference on Field and Service Robotics,* pp. 208–219, 1999.

6. Thrun S., et al., "Experiences with two deployed interactive tour-guide robots," *Proceedings of the International Conference on Field and Service Robotics*, pp. 37–42, 1999.

7. Torrie M., S. Veeramachaneni, B. Abbott, "Laser-based obstacle detection and avoidance system," *Proceedings of the SPIE Conference on Robotics and Semi-Robotic Ground Vehicle Technology*, pp. 2–7, 1998.

8. Kadam, V. G., Tamane, S. C., and Solanki, V. K., "Smart and connected cities through technologies." In *Big Data Analytics for Smart and Connected Cities*, pp. 1–24. IGI Global, Hershey, PA, 2019.

9. Sanju, D. D., A. Subramani and V. K. Solanki. "Smart City: IoT based prototype for parking monitoring & parking management system commanded by mobile app," *Second International Conference on Research in Intelligent and Computing in Engineering*, 2018.

10. Dhall, R., V. K. Solanki, "An IoT based predictive connected car maintenance approach," *International Journal of Interactive Multimedia and Artificial Intelligence*, 4(3), 2017.

11. Solanki, V. K., M. Venkatesan, S. Katiyar, "Conceptual model for smart cities for irrigation and highway lamps using IoT," *International Journal of Interactive Multimedia and Artificial Intelligence*, 4(3), 28–33, 2018.

12. Solanki, V., M. Venkatesan, S. Katiyar, "Think home: A smart home as digital ecosystem in circuits and systems," *Scientific Research*, 10(7), 2018.

13. Solanki, V. K., S. Katiyar, V. B. Semwal, P. Dewan, M. Venkatesan, N. Dey, "Advance automated module for smart and secure city," *ICISP, Procedia Computer Science*, 78, 367–374, 2015.

14. Semwal, V. B., P. Chakraborty, G. C. Nandi. "Less computationally intensive fuzzy logic (type-1)-based controller for humanoid push recovery," *Robotics and Autonomous Systems* 63, 122–135, 2015.

15. Semwal, V. B., K. Mondal, G. C. Nandi. "Robust and accurate feature selection for humanoid push recovery and classification: Deep learning approach," *Neural Computing and Applications*, 1–10, 2015.

16. Moore K. and M. Torrie, "*John Deere Final Report: Obstacle Detection and Avoidance for an Automated tractor,*" Tech. Rep, Center for Self-Organizing and Intelligent Systems, Utah State University, August. 1999.

17. Semwal, V. B., A. Bhushan, G. C. Nandi, "Study of Humanoid Push Recovery Based on Experiments," *Control, Automation, Robotics and Embedded Systems (CARE), 2013, International Conference on*. IEEE, 2013.

4

XBee and Internet of Robotic Things Based Worker Safety in Construction Sites

Rajesh Singh, Anita Gehlot, Divyanshu Gupta, Geeta Rana, Ravindra Sharma, and Shivani Agarwal

CONTENTS

4.1 Introduction

Worker safety on constructor sites is in the hands of supervisors. In recent days technology plays the role of supervisor in the construction site, supervising direct communication between different "smart things" such as sensors, smart phones, and databases, which offers a new angle on safety management using the Internet of Things [1]. Many accidents occur every day on construction sites; around the world, construction companies have implemented various types of construction safety measures to reduce the likelihood of accidents on sites. To reduce accident many companies have placed poster and warning symbols [2]. To protect workers on the construction site and reduce accidents in dangerous sites, this chapter proposes a smart technology for the system that gathers data of the construction site, sends the data to the cloud, and warns construction site labors who work within dangerous work sites. The Internet of Things (IoT) is used in this system for the storage and analysis of data. To detect and monitor workers on the construction sites, the system consists of three techniques: (1) long-range radio frequency, (2) high-db antennas for bidirectional communication, and (3) ultrasound waves of 10 Khz. Vehicles contain a transreceiver that communicates to the control. The model of the portable device contains a GPS module and a wake-up sensor. The wearable device contains a power-saver system: an automatic wake-up sensor that activates automatically when the worker moves, via when the smart device senses pressure on it; otherwise it remains in deep sleep. Because of this, the life of the battery increases up to twice as long. Furthermore, this chapter presents wireless nodes, which can be powered by PV cells that provide energy in the daytime, and this stored power can be used in the nighttime. They can be used indoors and outdoors [3]. In recent years, there has been improved management of construction safety using technologies in wireless sensor networks (WSNs), which are interlinks to the Internet of Things that communicate between different types of sensors and devices. One of the development sites enables sensors to sense and give readings of environmental conditions. One article discusses the use of BIM and WSN in a wireless system that helps the construction-site worker to continuously monitor safety parameters via a different sensor with a colored interface, which automatically informs the worker of any hazardous gas. Some wireless sensors are placed in hazardous underground conditions. With the help of a model like BIM, the construction site region is alerted and given alarm and warning signals when a dangerous condition happens. This system greatly enhances efficiency in construction safety management and provides an important reference information in rescue tasks [4]. Another article looks at the appropriateness of adopting different types of automated construction safety tools with regard to construction safety, costs, and benefits [5]. A further article discusses CAR (construction automation and robotics) with a framework to do the assessment of building construction projects. The research adopts the V model approach to convert the framework into an assessment method [6]. The smart sensor and smart protective equipment has been developed to improve workers' safety and protection on the site. Using these types of solutions helps to modify traditional work methods; production increases processes' complexity, and with the use of high dynamism, smart working environments were created [7]. Wearables for construction safety are an emerging technology. T here are some technical challenges in the wearables, such as battery life of the device, administration of wearables on a daily basis, and creating robust analytics [8]. In a road construction project, the system was tested and the results showed the system that could be used to improve construction safety. This system contains safety information, analysis, monitoring, and highlights the useful data [9]. Many smart sensor–based technologies have been used in construction site safety kits, which contain global positioning system technology, vision sensor, and wireless node of sensor. This chapter proposes a review of previous studies in this field to find the useful research gap for future research [10]. With the help of software interfaces and networks, authorization services can be accessed, with access provided by accessors. Smart sensor technology (SST) provides various security satisfaction and features; using an automated verification tool, we perform security analysis. This analysis demonstrates stability with a mathematical approach, which, with the help of experiments, shows security of network entities [11]. We define current trends in information and communications technology (ICT) publications related to information flow and also identify the typical technologies used to enable these communication modes [12]. Unmanned aerial vehicle (UAV) are helpful for hazardous workplaces, helping to manage safety of the worker in the construction site. UAVs also perform some dangerous tasks, helping to improve safety of the workplace. An article shows three uses of UAVs in construction sites: potential risk, risk analysis, and improved safety parameters in the construction site [13]. Another article presents a novel system that combines 3D body and hand position tracking that was developed to capture the movements of human construction workers. The literature has been carried out for concluding the details of different technologies used for the automation of construction sites [14–15]. An article discussed analysis of force sensitive resistor (FSR), which is placed in the helmet and responsible for igniting the vehicle [16,17]. Other articles discussed the automation and monitoring of construction sites [18,19]. One article discussed the tracking of goods using IoT and intelligent sensors in the transport vehicle [20]. Vrushali Kadam et al., Rohit Dhall et al., and V.K. Solanki et al. discussed the importance of smart technologies in the development of smart cities [21–23]. The various application of the Internet of Things in the field of maintenance, irrigation, highway lamps, and smart home monitoring [24–26] is also discussed.

4.2 Hardware Development

Figure 4.1 shows the block diagram of the system, which contains the large number of XBee smart helmets. Each helmet act as a node, with each node having transceiver that communicates among the rest; this is also called a personal area network. All these smart helmets are deployed at the construction site for the safety of the worker; all these helmets contain a transmitter that measures various parameters inside the construction site and directly send this data to the XBee local coordinator. This XBee local coordinator coordinates to all the helmets to collect data and processes these data. The XBee local coordinator is further connected to the main server through the Internet.

Smart helmets consist of many sensors that measure various parameters and give these data to the Arduino LilyPad (controller) smoke sensor, which detects harmful gases like CO_2, methane, and so on. Then it gives pulse to the Arduino, which calculates the intensity of harmful gases. The IR sensor and

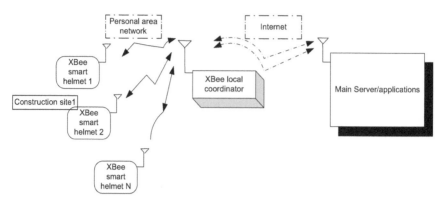

FIGURE 4.1
Generalized block diagram of the system.

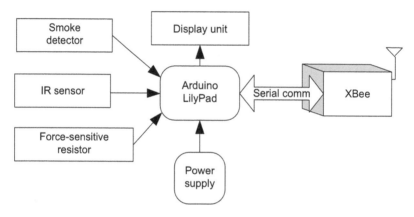

FIGURE 4.2
Helmet node end.

force sensor combined act when both sensors sense together; then only the smart helmet starts working; otherwise, it remains in deep sleep so that power is saved. The Arduino LilyPad is further connected to the XBee through serial communication; it transmits the status of the worker site. Figure 4.2 shows the block diagram of the helmet node end for construction workers.

Figure 4.3 shows the coordinator node at the local control room receiving data from different local nodes; these data are further processed by Arduino Uno and displayed at the display unit after processing of data. The Arduino Uno further forwards data to the application server through Wi-Fi communication. With the help of the Internet, we can access these readings and the working condition of the construction site.

Smart helmets are built with the latest technology for saving the life of workers; the smart helmet consists of many sensors like smoke sensor, flex sensor, IR sensor, and a display unit—all of these are connected by the

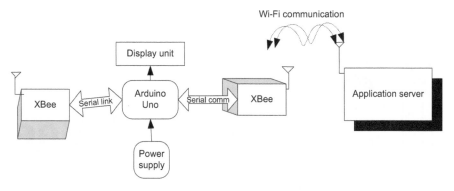

FIGURE 4.3
Coordinator device at local control room.

central controller Arduino LilyPad, which is further connected with XBee to transmit data wirelessly to the main server. When the sensor reaches the threshold value due to the working environment, then it gives data and an alarm signal to the application server, which is further connected to the XBee local coordinator, which is further linked with the Internet through the GSM module. All the data collected from the smart helmet are stored in the cloud and can be accessed throughout the world.

4.2.1 Flex Sensor

Flex sensors are flexible tubelike structure in which resistance of flux sensor changes when it bent this sensor are mainly used in glove. Suppose when we test the initial resistance reading of flex sensor, it is 30 K ohm; when it's sent to some angle, its resistance changes to 40–90 K. When the sensor comes to its initial condition, its resistance also comes back to 30 K.

4.2.2 Gas Sensor

Gas sensors are useful to measure and detect the presence of some harmful gases; they consist of variable resistance to adjust the sensitivity of that sensor. It can be used to detect leakage in home and industry, and it's very compact in size. It consists of three pins: one for Vcc and another for Vee and one for digital data out. It can detect propane, LPG (liquified petroleum gas), CH4, and alcohol.

Figure 4.4 shows the helmet node with construction workers. The flex sensor data Out pin is connected to the Arduino LilyPad pin A0, IR sensor output is connected to Arduino pin 2, Xbee transmit pin is connected to Arduino pin 1, and the XBee receive pin is connected to Arduino pin 0; the display is connected to pins 8–11 (4-bit data). Figure 4.5 shows the schematics of the coordinator device at the local control room; the Rx pin of the modem is connected to Arduino pin 1, XBee transmit pin is connected to pin 0, and LCD pin is connected to Arduino 8–11.

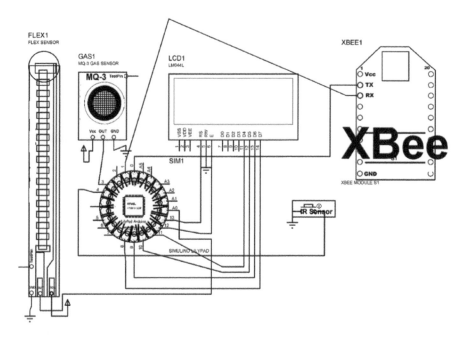

FIGURE 4.4
Schematics of helmet node end.

Table 4.1 shows the components list, which is used to design and interface circuits to design the system for the helmet node.

The following are the interfacing connections of Arduino LilyPad, FSR, LCD, XBee modem, and smoke detector in helmet node:

1. +5 V and GND pins of Arduino LilyPad are connected to +5 V and GND pins of power supply.
2. Pins 1 and pin 16 of LCD are connected to the GND of power supply.
3. Pins 2 and pin 15 of LCD are connected to +5 V of power supply.
4. Fixed legs of 10 K potentiometer (POT) is connected to +5 V and GND of power supply and the variable leg to pin 3 of LCD.
5. Pin 13, GND, and pin 12 of the Arduino LilyPad is connected to pin 4 (RS), pin 5 (RW), and pin 6 (E) of LCD.
6. Pin 11, pin 10, pin 9, and pin 8 of Arduino LilyPad is connected to pin 11 (D4), pin 12 (D5), pin 13 (D6), and pin 14 (D7) of LCD.
7. +Vcc, GND, and Out pins of FSR sensor are connected to +5 V, GND, and A0 pin of Arduino LilyPad.
8. +Vcc, GND and Out pins of MQ3 sensor are connected to +5 V, GND, and A1 pin of Arduino LilyPad.
9. +Vcc, RX, TX, and GND of XBee to the +5 V, TX, RX, and GND pins of Arduino LilyPad.

TABLE 4.1

Components Used for Helmet Node

Arduino LilyPad	1
+12 V/1A power supply	1
+12 V to +5 V converter	1
Extension of power supply breakout board	1
LCD20 * 4	1
LCD breakout board	1
2 LED with 330 ohm resistor	1
XBee modem	1
FSR sensor	1
MQ3 smoke detector	1
M-M connector	1
M-F connector	1
F-F connector	1

TABLE 4.2

Components Used for Coordinator Node

Arduino LilyPad	1
+12 V/1A power supply	1
+12 V to +5 V converter	1
Extension of power supply breakout board	1
LCD20 * 4	1
LCD breakout board	1
2 LED with 330 ohm resistor	1
XBee modem	1
M-M connector	1
M-F connector	1
F-F connector	1
NodeMCU	1

Table 4.2 shows the components list, which is used to design and interface with the circuit to design the system for coordinator node.

The following are the interfacing connections of Arduino LilyPad, LCD, and XBee modem in coordinator node:

1. +5 V and GND pins of the Arduino LilyPad are connected to the +5 V and GND pins of the power supply.
2. Pins 1 and pin 16 of the LCD are connected to the GND of power supply.
3. Pins 2 and pin 15 of the LCD are connected to +5 V of power supply.

4. Fixed legs of 10 K POT are connected to +5 V and GND of power supply and the variable leg to pin 3 of the LCD.

5. Pin 13, GND, and pin 12 of Arduino LilyPad are connected to pin 4 (RS), pin 5 (RW), and pin 6 (E) of LCD.

6. Pin 11, pin 10, pin 9, and pin 8 of the Arduino LilyPad are connected to pin 11 (D4), pin 12 (D5), pin 13 (D6) and pin 14 (D7) of LCD.

7. +Vcc, RX, and GND of XBee to the +5 V, TX and GND pins of Arduino LilyPad.

8. Connect Vin, GND, and TX pins of NodeMCU to Arduino Uno platform.

Figure 4.5 shows the schematics diagram of coordinator node, which will receive the data from the smart helmet regarding wearing the helmet or not.

Figures 4.6 and 4.7 show the CAD design of the helmet, which shows the placement of sensors in the helmet to sense the presence of wearing the helmet.

Table 4.3 shows the enabling technologies for the Internet of Things.

Table 4.4 shows the IoT-supported platform through which to send the required data to the cloud.

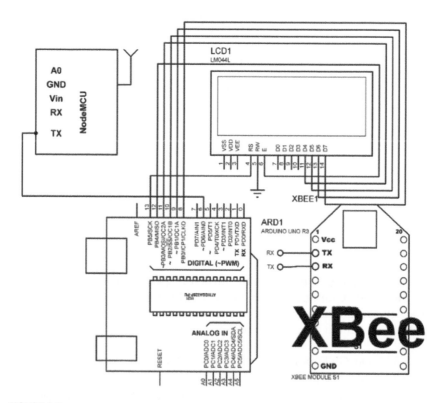

FIGURE 4.5
Schematics diagram of coordinator device.

Gas sensor   Ir sensor

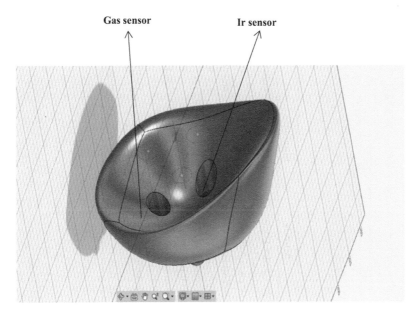

FIGURE 4.6
Side view of smart helmet.

Xbee transmitter

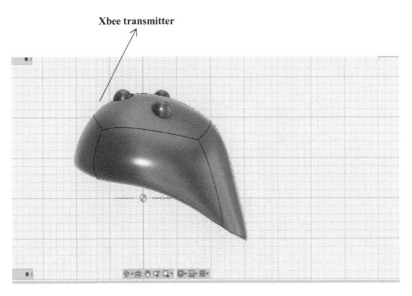

FIGURE 4.7
Top view of smart helmet.

TABLE 4.3

Enable Technologies for IoT

Parameters	Wi-Fi	WiMAX	LRWPAN	Mobile	Bluetooth	LoRA
Standard	IEEE802.11a/c/b/d/g/n	IEEE802.16	IEEE802.15.4 XBee	2G GSM, CDMA 3G-UMTS, CDMA2000, $G LTE	IEEE802.15.1	LoRA WAN R1.0
Frequency band	5–60 GHz	2–66 GHz	868/915 MHz, 2.4 GHz	865 MHz, 2.4 GHz	2.4 GHz	868/900 MHz
Data rate	1–6.75 Mb/s	1 Mb/s–1 Gb/s (Fixed) 50–100 Mb/s (mobile)	40–250 kb/s	2G:50–100 kb/s 3G:200 kb/s 4G:0.1–1 Gb/s	1–24 Mb/s	0.3–50 Kb/s
Transmission range	20–100 m	<50 Km	10–20 m	Entire cellular area	8–10 m	<30 Km
Energy consumption	High	Medium	Low	Medium	Low	Very Low
Cost	High	High	Low	Medium	Low	High

TABLE 4.4

IoT Supported Platform

Parameters	Arduino Uno	Arduino Yun	Intel Galileo gen2	Intel Edison	Beagle Bone Black	Electric IMP003	Raspberry Pi	ARM mbed NXP LPC1768	Telos B
Processor	Atmega 328p	Atmega32U4 and Atheros AR9331	Intel Quark SOC X1000	Intel Quark SOC X1000	Sitara AM3358BZC Z100	ARM cortex M4F	Broadcom BCM2835 SOC based ARM11	ARM cortex M3	MSP430
Operating Voltage	5 V	5 V, 3 V	5 V	3.3 V	3.3 V	3.3 V	5 V	5 V	3–36 V
Clock speed	16 MHz	16,400 MHz	400 MHz	100 MHz	1 GHz	320 MHz	700 MHz	96 MHz	8 MHz
Bus width	8	32	32	32	32	32	32	32	16
System memory	2 kB	2 Kb, 5 Kb, 64 Kb	256 MB	1 GB	512 MB	120 Kb	512 MB	32 Kb	10 Kb
Flash memory	32 Kb	32 Kb, 16 Mb	8 Mb	4 GB	4 GB	4 Mb	–	512 Kb	48 Kb
EEPROM	1Kb	1Kb	8Kb	–	–	–	–	–	–
Development environment	Arduino IDE	Arduino IDE	Arduino IDE	Arduino IDE, Eclipse, Intel XDK	Debian, Android, Ubuntu, Cloud9 IDE	Electric ImpIDE	NooBS	C/C++ SDK, Online compiler	Eclipse IDE
Programming language	Wiring	wiring	Wiring	Wiring, C, C++, Node JS, HTML5	C, C++, Python, Perl, Ruby, JAVA, Node8	squirrel	C, C++, Python, Ruby, JAVA	C, C++	C, NesC
I/O connectivity	SPI, I2C, UART, GPIO	SPI, I2C, UART, GPIO	SPI, I2C, UART, GPIO	SPI, I2C, UART, GPIO, I2S	SPI, I2C, UART, GPIO, McASP	SPI, I2C, UART, GPIO	SPI, DSI, UART, SDIO, CSI, GPIO	SPI, I2C, GPIO, CAN	USB serial, GPIO
Communication standard	IEEE 802.11 b/g/n, IEEE 802.15.4, 433RF, BLE4.0, Ethernet, Serial	IEEE 802.11 b/g/n, IEEE 802.15.4, 433RF, BLE4.0, Ethernet, Serial	IEEE 802.11 b/g/n, IEEE 802.15.4, 433RF, BLE4.0, Ethernet, Serial	IEEE 802.11 b/g/n, IEEE 802.15.4, 433RF, BLE4.0, Ethernet, Serial	IEEE 802.11 b/g/n, IEEE 802.15.4, 433RF, BLE4.0, Ethernet, Serial	IEEE 802.11 b/g/n, IEEE 802.15.4, 433RF, BLE4.0, Ethernet, Serial	IEEE 802.11 b/g/n, IEEE 802.15.4, 433RF, BLE4.0, Ethernet, Serial	IEEE 802.11 b/g/n, IEEE 802.15.4, 433RF, BLE4.0, Ethernet, Serial	CC2420

IoT-enabled systems can be virtual, physical, or hybrid in nature. They consist of the collection of physical things, sensors, actuators cloud services, specific IoT protocols, communication layers, users, developers, and the enterprise layer. Table 4.4 shows the platform through which to do the processing of sensors and actuators. It contains a hardware platform like Arduino Uno, Intel Edison, Intel Galileo, ARM mbed NXP LPC1768, Beagle Bone Black, Electric Imp 003 Raspberry Pi BC, Arduino Yun, Gen 2, and TelosB. The comparison of several units have been categorized based on parameters like Flash memory, processing unit, speed of clock, operating voltage, memory of the system, integrated drive electronics (IDE), languages to support the programming, input and output connectivity, type of processors, and bus width. The given processing unit supports the development of applications like waste management, home automation, construction, and others.

4.3 Software Development

The following program code of the helmet node is compiled in Arduino IDE and uploaded in LilyPad Arduino.

```
#include <LiquidCrystal.h>// add the liquid crystal display
library
const int rs = 13, en = 12, d4 = 11, d5 = 10, d6 = 9, d7 = 8;
LiquidCrystalDISPLAY(rs, en, d4, d5, d6, d7);// assign pins to
Arduino for LCD
int FSR_sensor_pin=A0;// assume integer to read analog sensor
int MQ3_sensor_pin=A1;// assume integer to read analog sensor
void setup()
{
DISPLAY.begin(20, 4);// initialize liquid crystal display
DISPLAY.print("SMART Helmet system");// print string on liquid
crystal display
Serial.begin(9600);// Initialize serial 9600 - 8- N- 1
delay(2000);// provide delay of 2sec
lcd.clear();// clear liquid crystal display
 }
void loop()
{
FSR_level=analogRead(FSR_sensor_pin);// read analog pin at
level FSR_sensor_pin
MQ3_level=analogRead(MQ3_sensor_pin);// read analog pin at
level MQ3_sensor_pin
DISPLAY.setCursor(0, 0);// set cursor of liquid crystal
display
DISPLAY.print("FSR_level:");// print string on liquid crystal
display
```

```
DISPLAY.setCursor(0, 1);//set cursor of liquid crystal display
DISPLAY.print(FSR_level);// print int of leveled FSR_level on
liquid crystal display
DISPLAY.setCursor(0, 2);//set cursor of liquid crystal display
DISPLAY.print("MQ3_level:");// print string on liquid crystal
display
DISPLAY.setCursor(0, 3);//set cursor of liquid crystal display
DISPLAY.print(MQ3_level);//print int of leveled FSR_level on
liquid crystal display
Serial.print(FSR_level);// print FSR_level integer on the serial
Serial.print(",");//print string on the serial
Serial.print(MQ3_level);//// print MQ3_level integer on the
serial
Serial.print('\n');//print string on the serial
delay(50);// provide delay of 50 mSec
}
```

The following program code of the coordinator node is compiled in Arduino IDE and uploaded in LilyPad Arduino.

```
int X, Y;// let assume int X and Y
String FSR_MQ _STRING = "";// let assume a string
#include <LiquidCrystal.h>// add library of liquid crystal
display
const int rs = 13, en = 12, d4 = 11, d5 = 10, d6 = 9, d7 = 8;//
assign pins to Arduino of liquid crystal display
LiquidCrystalDISPLAY(rs, en, d4, d5, d6, d7);//assign pins to
Arduino of liquid crystal display
void setup()
{
DISPLAY.begin(20, 4);// initialize liquid crystal display
20 x 4
Serial.begin(9600);// Initialize serial 9600 - 8- N- 1
DISPLAY.print("SMART Helmet system");// print string on LCD at
leveled "SMART Helmet system"
delay(2000);// provide delay of 2 sec
lcd.clear();// clear the LCD
}
void loop()
{
serialEvent_arduino();// call the function to read the serial
value from Arduino TX
DISPLAY.setCursor(0, 0);// set cursor on liquid crystal
display to print the data
DISPLAY.print("FSR_level:");// print string "FSR_level:"on
liquid crystal display
DISPLAY.setCursor(0, 1);// set cursor on liquid crystal
display to print the data
DISPLAY.print(X);// print int value of X on liquid crystal
display
```

```
DISPLAY.setCursor(0, 2);// set cursor on liquid crystal
display to print the data
DISPLAY.print("MQ3_level:");//print string "MQ3_level:"on
liquid crystal display
DISPLAY.setCursor(0, 3);//set cursor on liquid crystal display
to print the data
DISPLAY.print(Y);//print int value of Y on liquid crystal
display
Serial.print(X);// print int value of Y on serial pin TX of
Arduino
Serial.print(",");//print string on serial pin TX of Arduino
Serial.print(Y);//print int value of Y on serial pin TX of
Arduino
Serial.print('\n');// print new line char '\n' serial pin TX
of Arduino
delay(50);// provide delay of 50mSec
}

void serialEvent_arduino() // function to receive the serial
data from Arduino TX
{
  while (Serial.available()>0)// check the serial data greater
than 0
  {
FSR_MQ_STRING= Serial.readStringUntil('\n');// Get serial
input
X=(((FSR_MQ_STRING[0]-48)*1000)+((FSR_MQ_STRING[1]-
48)*100)+((FSR_MQ_STRING[2]-48)*10)+
((FSR_MQ_STRING[3]-48)*1)));
Y=(((FSR_MQ_STRING[5]-48)*1000)+((FSR_MQ_STRING[6]-
48)*100)+((FSR_MQ_STRING[7]-48)*10)+ ((FSR_MQ_
STRING[8]-48)*1)));// convert the string into Decimal format
}
FSR_MQ_STRING = "";// make empty the string
delay(100);// provide delay 100 mSec
  }
```

The following program code of the coordinator node is compiled in
Arduino IDE and uploaded in NodeMCU if it realizes the application using
Cayenne.

```
#define CAYENNE_PRINT Serial// define Cayenne serial print
#include <CayenneMQTTESP8266.h>// add library of Cayenne
MQTT8266
charWiFi_ID[] = "ESPServer_RAJ";// ID of Hot Spot
char WiFi_Password[] = "RAJ@12345";// password of Hot Spot

char US[] = "fac81bb0-7283-11e7-85a3-9540e9f7b5aa";// user
token
```

```
char PS[] = "3745eb389f4e035711428158f7cdc1adc0475946";//
client password
char CI[] = "386b86f0-7284-11e7-b0bc-87cd67a1f8c7";// client
token

unsigned long Lmilli = 0;// assume unsigned long
int FSR, MQ3;// assign int to FSR, MQ3
String inputString_NODEMCU = "";// assign string to store
serial data
void setup()
{
Serial.begin(9600);// // Initialize serial 9600 - 8- N- 1
Cayenne.begin(UN, PS, CI, WiFi_ID, WiFi_Password);// initialize
the cayenne server
}

void loop()
{
Cayenne.loop();// start cayenne loop
serialEvent_NodeMCU(); call serial event to receive the data
of Arduino using NodeMCU
if (millis() - Lmilli> 10000)// if loop
  {
Lmilli = millis();
Cayenne.virtualWrite(0, X);// virual write the data of X
variable to cayenne virtual field 1
Cayenne.virtualWrite(2, Y);// // virual write the data of Y
variable to cayenne virtual field 2
  }
}

void serialEvent_NodeMCU() // Serial event function to receive
the data from arduino
{
  while (Serial.available()>0)// accept serial data which is
greater than 0
  {
FSR_MQ_STRING= Serial.readStringUntil('\n');// Get serial
input
FSR=(((FSR_MQ _STRING[0]-48)*1000)+((FSR_MQ _STRING[1]-
48)*100)+((FSR_MQ _STRING[2]-48)*10)+ ((FSR_MQ
_STRING[3]-48)*1)));
MQ3=(((FSR_MQ _STRING[5]-48)*1000)+((FSR_MQ _STRING[6]-
48)*100)+((FSR_MQ _STRING[7]-48)*10)+ ((FSR_MQ
_STRING[8]-48)*1)));//
  }
FSR_MQ _STRING = "";// make empty the given string
delay(100);// provide delay of 100 m sec
}
```

The following program code of the coordinator node is compiled in Arduino IDE and uploaded in NodeMCU, if it realizes the application using Blynk.

```
#define BLYNK_PRINT Serial // define blynk lib
#include <ESP8266WiFi.h>// add Wi-Fi header
#include <BlynkSimpleEsp8266.h>// add blynk header lib

#include <LiquidCrystal.h>// add liquid crystal lib
LiquidCrystallcd(D0, D1, D2, D3, D4, D5);// add pins
char auth[] = "7d5b9ec1f13e4e29b34250a3c72f08e9";// add token
generated by application
char ssid[] = "ESPServer_RAJ";// user name of hot spot
char pass[] = "XXX@12345";// password of hot spot
BlynkTimer timer;// add blynk timer
int FSR, MQ3;// assign integer
String FSR_MQ_STRING;// assign string
void READ_SENSOR()// function to read sensors
{
READ_SENSOR_NodeMCU();// function to receive data for
coordinator node.
Blynk.virtualWrite(V0, FSR);// write FSR value on virtual pin
V0
Blynk.virtualWrite(V1, MQ3);// write FSR value on virtual pin
V1
lcd.setCursor(0,0);// set cursor of LCD
lcd.print("FSR:");// print string on LCD
lcd.setCursor(0,1);// set cursor of LCD
lcd.print(FSR);// print int value of FSR
lcd.setCursor(0,2);// set cursor on LCD
lcd.print("MQ3:");// print string on LCD
lcd.setCursor(0,3);// set cursor of LCD
lcd.print(MQ3);// print int value of MQ3
}

void setup()
{
Serial.begin(9600);// begin serial communication
lcd.begin(20, 4);// initialize LCD
Blynk.begin(auth, ssid, pass);// initialize Blynk APP
timer.setInterval(1000L, READ_SENSOR);//// set timer delay
}

void loop()
{
Blynk.run();// call Blynk run function
timer.run();// InitializeBlynkTimer
}

void READ_SENSOR_NodeMCU () // Serial event function to
receive the data from arduino
```

```
{
while (Serial.available ()>0)// accept serial data which is
greater than 0
{
FSR_MQ_STRING= Serial.readStringUntil('\n');// Get serial
input
FSR=(((FSR_MQ _STRING[0]-48)*1000)+((FSR_MQ _STRING[1]-
48)*100)+((FSR_MQ _STRING[2]-48)*10)+ ((FSR_MQ
_STRING[3]-48)*1)));
MQ3=(((FSR_MQ _STRING[5]-48)*1000)+((FSR_MQ _STRING[6]-
48)*100)+((FSR_MQ _STRING[7]-48)*10)+ ((FSR_MQ
_STRING[8]-48)*1)));//
}
FSR_MQ _STRING = "";// make empty the given string
delay(100);// provide delay of 100 m sec
}
```

Figure 4.8 shows the flowchart of the helmet node with construction workers to log safety and attendance on a cloud server. Figures 4.9 and 4.10 show the flowchart of the coordinator device at a local server that communicates with the main server using NodeMCU.

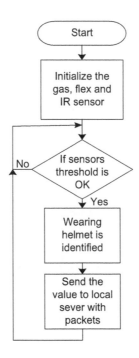

FIGURE 4.8
Flow chart of helmet Node.

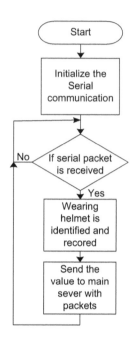

FIGURE 4.9
Flow chart of coordinator.

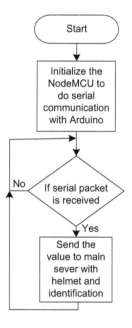

FIGURE 4.10
Flow chart for NodeMCU.

4.4 IoT Implementation

The communication models of the Internet of Things are request-response communication, publish-subscribe communication, push-pull communication, and exclusive pair communication.

Request-response is a communication model in which the client sends requests to the server and the server responds to the requests. When the server receives a request, it decides how to respond, fetches the data, retrieves resource representations, prepares the response, and then sends the response to the client.

Exclusive pair is a bidirectional, full duplex communication model that uses a persistent connection between client and server. Once the connection is set up, it remains open until the client sends the request to close the connection. Client and server can send the messages to each other after connection setup. Exclusive pair communication model and the server are aware of all the open connections.

The most popular IoT communication application programmable interfaces (APIs) are representational state transfer (REST) and WebSocket.

The representational state transfer (REST), or RESTful Web services, is the way to provide interoperability between computer systems that are connected to the Internet. Such types of Web services provide requests to the system to access and manipulate the textual representations of Web resources. The system uses the uniform and predefined set of stateless operations, which exposes the arbitrary sets of operations such as web service definition language (WSDL) and simple object access protocol (SOAP). The Web resources are already available in the World Wide Web (www) as documents or files identified by their URLs. The RESTful Web service and the resource URI then divide a response in XML, HTML, JSON, or some other defined format. The HTTP format uses the create, read, update and delete (CRUD) HTTP methods with commands GET, POST, PUT, DELETE, and so on.

On the other hand, WebSocket is a computer communications protocol that provides full duplex communication channels using a single transmission control protocol (TCP) connection. The WebSocket protocol was standardized by the IETF as RFC 6455 in 2011. The WebSocket application program interface (API) in Web IDL was standardized by the W3C. It is a different TCP from HTTP. It is a two-way (bi-directional) conversation and able to establish communication between the client and the server.

A Level 1 IoT system has a single node/device that performs sensing and/or actuation, stores data, performs analysis, and hosts the application. The system is suitable for modeling low-cost and low-complexity solutions in which the data involved is not big and the analysis requirements are not computationally intensive. An example is **home automation,** whereby a single node allows controlling the lights and appliances in a

home remotely. A Level 2 IoT system has a single node/device that performs sensing and/or actuation and local analysis. The data is stored in
the cloud and the application is cloud based. The system is suitable when
the data involved is big. However, the primary analysis requirement is not
computationally intensive and can be done locally. An example is smart
irrigation. The system has a single node that monitors soil moisture level
and controls the irrigation system. A Level 3 IoT system has a single node.
Data is stored and analyzed in the cloud and the application is cloud based.
This type of system is suitable when data involved is big and the analysis
requirements are computationally intensive. An example is tracking the
workers for wearing their helmet. The system consists of a single node (for
a helmet) that monitors the safety and attendance for construction workers.

A Level 4 IoT system has multiple nodes that perform local analysis.
Data is stored in the cloud and the application is cloud based. The local
and cloud-based observer nodes can subscribe to and receive information
collected in the cloud from IoT devices. Observer nodes can process information and use it for various applications. However, observer nodes do
not perform any control functions. This type of system is suitable when
lot of nodes are involves in the process, the data is big, and the analysis
requirements are computationally intensive. City noise monitoring is one
best example in Level 4. The system consists of multiple nodes placed in
different locations for monitoring noise levels in an area. A Level 5 IoT
system has a large number of nodes and end nodes and one coordinator.
The end nodes perform sensing and/or actuation. The coordinator node
collects data from the end nodes and sends it to the cloud. Data is stored
and analyzed in the cloud and the application is cloud based. The system
is suitable for solutions based on WSN, in which the data involved is big
and the analysis requirements are computationally intensive. An example
is forest fire detection. The system consists of various nodes placed in different locations of the forest to monitor the temperature, humidity, and
CO2 levels. A Level 6 IoT system has multiple independent end nodes
that perform sensing and/ or actuation and send data to the cloud. Data
is the stored in the cloud and the application is cloud based. The analytics
component analyzes the data and stores the results in the cloud database.
The results are visualized with the cloud-based application. The centralized controller is aware of the status of all the end nodes and sends control
commands to the nodes. An example is weather monitoring. The system
consists of multiple nodes that are placed in different locations for monitoring temperature, humidity pressure, radiation, and wind speed. The
end nodes are equipped with various sensors. The end nodes send the
data to the cloud in real time using WebSocket service. The data is stored
in a cloud-based server. The analysis of data is done in the cloud to aggregate the data and make the predictions. A cloud-based application is used
for visualizing the data.

4.5 Internet of Things Design Methodology

The steps to design an IoT-related system are as follows:

Step 1—**Purpose and requirements specification:** The first step in IoT system design methodology is to define the purpose and requirements of the system. In this step, the system purpose, behavior, and requirements (e.g., data collection requirements, data analysis requirements, system management requirements, data privacy and security requirements, and user interface requirements) are captured.

Applying this to our example of a smart IoT-enabled robot system, the purpose and requirements for the system may be described as follows:

a. Purpose: A robot control system that allows controlling the direction of robots using the Internet of Things.

b. Behavior: The switches on the app provide the data, which is either 10, 20, 30, and so on, serially. According to received data, the command for the robot is to move forward, reverse, left, right, and stop.

c. System management requirement: The system should provide remote monitoring and control functions.

d. Data analysis requirement: The system should perform local analysis of the data.

e. Application deployment requirement: The application should be deployed locally on the device, but should be accessible remotely.

Step 2—**Process specifications:** The second step in the IoT design methodology is to define the process specification. In this step, the use cases of the IoT system are formally described based on and derived from the purpose and requirement specifications.

Step 3—**Domain model specification:** The third step in the design methodology is to define the domain model. The domain model describes the main concepts, entities, and objects in the domain of the IoT system to be designed.

Step 4—**Information model specification:** The fourth step in the IoT design methodology is to define the information model. The information model defines the structure of all the information in the IoT system; for example, attributes of virtual entities, and relations. The information model does not describe the specifics of how the information is represented or stored.

Step 5—Service specifications: The fifth step in the design methodology is to define the service specifications. Service specifications define the services in the IoT system, service types, service inputs/output, service endpoints, service schedules, service preconditions, and service effects.

Step 6: The sixth step in the IoT methodology is to define the IoT level for the system.

Step 7—Functional groups: The seventh step in the design methodology is to define the functional view. The functional view defines the functions of the IoT system grouped into various functional groups. Each functional group either provides functionalities for interacting with instances of the concepts defined in the domain or provides information related to these concepts.

Step 8—Operational view specifications: The eighth step in the IoT design methodology is to define the operational view specifications. In this step, various options pertaining to the IoT system deployment and operation are defined, such as service hosting options, storage options, device options, and application-hosting options.

An operational view example mapping functional groups to operational view specifications for Robot Control using IoT system is as follows:

- **Devices:** Computing device (NodeMCU), Motor driver L293D, Two DC geared motors.
- **Communication APIs:** REST APIs.
- **Communication protocols:** Link layer-802.11, Network layer-IPv4/IPv6, Transport layer-TCP
- **Application:** HTTP
- **Services:**
  - Controller service: Hosted on device, implemented in Python, and run as a native service.
  - Mode Service: RESTful Web service, hosted on device and implemented with Django-REST framework.
  - State service: RESTful Web service, hosted on device and implemented with Django-REST framework.
  - **Application:**
    - Web application—Django Web application
    - Application server—Django app server
    - Database server—MySQL
    - Security

- Authentication: Web app, database
- Authorization: Web app, database
- Management
- Application management: Django app management
- Database management: MySQL DB management
- Device management: ESP8266 device management

4.6 Application Development

The following steps are used to add NodeMCU in the Cayenne cloud:

1. Install the Arduino IDE and add Cayenne MQTT Library to Arduino IDE.
2. Install the ESP8266 board package to Arduino IDE.
3. Install required USB driver on computer to program the ESP8266.
4. Connect the ESP8266 to PC/Mac via data-capable USB cable.
5. In the Arduino IDE, go to the **Tools** menu, select the **Board**, and now the **Port** ESP8266 is connected to.
6. Use the MQTT username, MQTT password, and client ID as well as ssid [] and wifiPassord [] in the Arduino IDE to write code.
7. Burn the code in Arduino and NodeMCU, then window will open.

The other platform to realize IoT is Blynk. Blynk is a platform based on mobile application and supports to iOS and Android apps. The application is control-using the Arduino, Raspberry Pi, and NodeMCU. The Blynk application, using widgets and by dragging and dropping, makes the graphical interface. Instead, it is supporting hardware of your choice. We can make Arduino or Raspberry Pi over the Internet by using Wi-Fi, Ethernet, or ESP8266 chip.

A platform like Blynk is for iOS, and the Android platform is used to design mobile apps. To design the app, download the latest Blynk library from https://github.com/blynkkk/blynk-library/releases/latest. The mobile app can easily be designed just by dragging and dropping widgets on the provided space. Tutorials can be downloaded from http://www.blynk.cc.

The following are the steps to develop Blynk APP:

Step 1: Download and install the Blynk app for your mobile Android or iPhone from http://www.blynk.cc/getting-started/

Step 2: Create a Blynk account.

Step 3: Create a new project.

Step 4: Auth token is a unique identifier that will be received on the email address user, provided at the time of creating an account. Save this token, as this is required to copy in the main program of the receiver section.

Step 5: Select the device to which the smart phone needs to communicate; e.g., ESP8266 (NodeMCU).

Step 6: Open the widget box and select the components required for the project. For this project five buttons are selected.

Step 7: Tap on the widget to get its settings and select virtual terminals as V1, V2 for each button,, which needs to be defined later on the program.

Step 8: After completing the widget settings, run the project.

4.7 Result and Discussion

The three main features of the system are emergency alarming, various parameter monitoring, and positioning of the worker and the affected place. By combining three technologies these can be achieved, and they are the (1) XBee, (2) smoke sensor, and (3) flux sensor. The flux sensor and IR sensor module are helpful for waking up the smart helmet module from the standby mode. For harmful gas measurement, the high-resolution capacity of the smoke sensor is used.

Figure 4.11 shows the data recorded by sensors at helmet node by the Cayenne server on the cloud. The IR sensor gives a digital High (1) when it

FIGURE 4.11
Snapshot of the data record of helmet node.

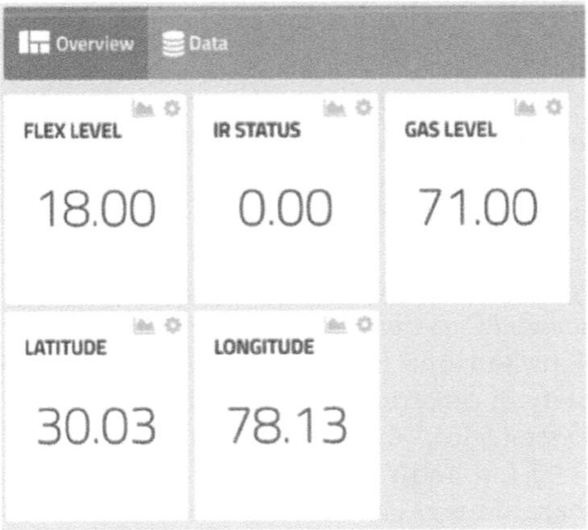

FIGURE 4.12
Snapshot of the data record of helmet node.

senses the object. The flex level is 33 and the gas contents level is 98, if the controller resolution is of 10 bit ADC. The latitude and longitude of the helmet node are also recorded for location measurement.

Figure 4.12 shows the data recorded by sensors at helmet nodes by the Cayenne server on the cloud. The IR sensor gives digital Low (0) when it does not sense the object/helmet. The flex level is 18 and the gas contents level is 71, if the controller resolution is of a 10-bit ADC. The latitude and longitude of the helmet node are also recorded for location measurement.

4.8 Conclusion and Future Scope

A safety solution for preventing accidents in construction sites was presented. With the use of advanced design techniques in the model, the electric circuitry, and system levels, this chapter represents IoT-based automatic systems, power saving, and real-time sensing systems: (1) accidents prevention system, and (2) alert system for reducing accidents. This type of system is designed to be used at construction sites, operates in real-time conditions, and monitors every change in the construction site. Multiple tests were done to check the working of this system. Future improvements on the smart helmet may include adding more sensors, based on the working conditions for improving the safety scale. With the help of this, we are able to continuously monitor the health of the workers on the construction site. In addition, a future improvement plan is the extension of the Internet of Things and extension of the range of XBee so that we can spread to a larger area. With the help of this, workers will work safely in the dangerous environment.

References

1. Li, R. Y. M. "Smart working environments using the Internet of Things and construction site safety." *An Economic Analysis on Automated Construction Safety.* Springer, Singapore (2018): 137–153.
2. Li, R. Y. M. "Turning the tide in the construction industry: From traditional construction safety measures to an innovative automated approach." *An Economic Analysis on Automated Construction Safety.* Springer, Singapore, (2018): 1–22.
3. Kanan, R., O. Elhassan, and R. Bensalem. "An IoT-based autonomous system for workers' safety in construction sites with real-time alarming, monitoring, and positioning strategies." *Automation in Construction* 88 (2018): 73–86.
4. Cheung, W. F., T. H. Lin, and Y. C. Lin. "A real-time construction safety monitoring system for hazardous gas integrating wireless sensor network and building information modeling technologies." *Sensors* 18.2 (2018): 436.
5. Li, R. Y. M. "RAND appropriateness study in regard to automated construction safety: A global perspective." *An Economic Analysis on Automated Construction Safety.* Springer, Singapore, (2018): 155–173.
6. Pan, M. et al. "A framework of indicators for assessing construction automation and robotics in the sustainability context." *Journal of Cleaner Production* 182 (2018): 82–95.
7. Podgórski, D. et al. "Towards a conceptual framework of OSH risk management in smart working environments based on smart PPE, ambient intelligence and the Internet of Things technologies." *International Journal of Occupational Safety and Ergonomics* 23.1 (2017): 1–20.
8. Korman, D. B. and A. Zulps. "Enhancing construction safety using wearable technology." *ASSE Professional Development Conference and Exposition.* American Society of Safety Engineers, Red Hook: Curran Associates, (2017).
9. Zou, P. X. W. et al. "Cloud-based safety information and communication system in infrastructure construction." *Safety Science* 98 (2017): 50–69.
10. Zhang, M., T. Cao, and X. Zhao. "Applying sensor-based technology to improve construction safety management." *Sensors* 17.8 (2017): 1841.
11. Kim, H. et al. "A toolkit for construction of authorization service infrastructure for the Internet of Things." *Proceedings of the Second International Conference on Internet-of-Things Design and Implementation.* New York: ACM, (2017).
12. Alsafouri, S., and S. K. Ayer. "Review of ICT implementations for facilitating information flow between virtual models and construction project sites." *Automation in Construction* 86 (2018): 176–189.
13. Howard, J., V. Murashov, and C. M. Branche. "Unmanned aerial vehicles in construction and worker safety." *American Journal of Industrial Medicine* 61.1 (2018): 3–10.
14. Kurien, M. et al. "Real-time simulation of construction workers using combined human body and hand tracking for robotic construction worker system." *Automation in Construction* 86 (2018): 125–137.
15. Rajesh, S., A. Gehlot, V. P. Singh, V. Garg, S. Kumar, S. Choudhury, R. Pachauri "Role of automation in construction industries: A review." *Journal of Engineering Technology* 6.2 (2017): 799–831.

16. Gehlot, A. et al. "Development and analysis of FSR and RFID based authentication system." *Proceeding of International Conference on Intelligent Communication, Control and Devices*. Springer, Singapore, (2017).

17. Gehlot, A. et al. "Design and analysis of controller for vehicle ignition using FPGA." *Indian Journal of Science and Technology* 9(43) (2016).

18. Agarwal, A. et al. "Arduino- and IoT-based tools and inventory tracking system in construction sites." *Intelligent Communication, Control and Devices*. Springer, Singapore, (2018): 1677–1686.

19. Gaba, A. et al. "An approach to monitor construction site based on radio frequency identification and Internet of Things." *Intelligent Communication, Control and Devices*. Springer, Singapore, (2018): 1629–1639.

20. Singh, A. K. et al. "Consistent tracking of goods throughout the transportation chain based on Internet of Things and wireless sensor network." *Journal of Communication Engineering & Systems* 7.3 (2018): 36–44.

21. Vrushali, K., S. Tamane, V. Solanki "Smart and connected cities through technologies," *IGI-Global*, doi:10.4018/978-1-5225-6207-8.ch001.

22. Durga Devi Sanju, A. Subramani and V. K. Solanki "Smart city: IoT based prototype for parking monitoring & parking management system commanded by mobile app" in *Second International Conference on Research in Intelligent and Computing in Engineering*, (2017).

23. Solanki, V. K., S. Katiyar, V. B. Semwal, P. Dewan, M. Venkatesan, N. Dey "Advance Automated Module for Smart and Secure City" in ICISP-15, organised by G.H.Raisoni College of Engineering & Information Technology, Nagpur, on 11–12 December (2015), published by Procedia Computer Science, Elsevier, ISSN 1877-0509et.

24. Rohit, D., and V. K. Solanki, "An IoT based predictive connected car maintenance approach" *International Journal of Interactive Multimedia and Artificial Intelligence*, 4.3 (2017), (ISSN 1989–1660).

25. Solanki, V. K., V. Muthusamy & S. Katiyar "Conceptual model for smart cities for irrigation and highway lamps using IoT," *International Journal of Interactive Multimedia and Artificial Intelligence*, (2018), (ISSN 1989–1660).

26. Solanki, V. K., V. Muthusamy & S. Katiyar "Think home: A smart home as digital ecosystem in circuits and systems," *Scientific Research Publishing Inc.* 10.7(2018), ISSN 2153–1293.

5

Contribution of IoT and Big Data in Modern Health Care Applications in Smart City

Mamata Rath and Vijender Kumar Solanki

CONTENTS

5.1 Introduction

It can be seen that the way we use driving technology is changing; a thrilling change is shaping the world from isolated systems to a pervasive method for an internet of "things." These things are outfitted for speaking with each other by sending data which contain critical information. In any case, this new world in view of the Internet contains different challenges from the persepective of security and insurance. The Internet of Things is described as a composed interconnection of contraptions in general use that are oftentimes furnished with an all-inclusive framework. The IoT relies upon treatment of immense measures of data, with a particular important objective to organize the data. Close to physical articles, the IoT [1,2] is made out of embedded programming, devices and sensors. This empowers articles to be controlled remotely by methods for the related orchestrated establishment and supports facilitated consolidation between the physical world and PC correspondence frameworks. Thus, it out and out adds to upgrade generosity, exactness, efficiency, and money-related advantages. This is the reason IoT [3,4] has been extensively associated in different applications, for instance, condition checking, essentialness organization, building computerization, transportation, thus on. The IoT [5,6] is a sort of structure using an arrangement of associated devices, sensors and looking at applications that work over the internet. It grasps particular taking care of and correspondence structures, progressions and layout methodology formed on their goal ?. The work uncovered in ?, sees that the IoT [7] thought has three essential properties: exhaustive care, strong transmission and adroit taking care of. At first in IoT [8,9] structure, the thorough recognizing using RFID, sensors and M2M terminal is to get information of smart dissents in the zone after some time. Additionally, strong transmission is to guarantee continuously the security, correspondence, coordinating and encryption with high exactness and with different frameworks traditions. Thirdly, clever getting ready which depends upon sharp preparing headways.

WoT (Web of Things) is an growing technology that combines and connects smart devices into Web by reusing and extending Web standards. While Web technology makes the developers' job easier, it faces security, management and efficiency challenges. With the rapid development of big data analytics [10], mobile computing, Internet of Things, cloud computing, and social networking, cyberspace has expanded to a cross-fused and ubiquitous space made up of human beings, things, and information. Internet applications have evolved from Web 1.0 to Web 2.0 and Web 3.0, and web information has seen an explosive growth, which is strongly promoting the advent of a global era of big data.

5.2 Relevance of IoT Integrated with WoT in Wireless Systems

The security is described as a plan of frameworks to shield unstable data from exposed strikes and to guarantee classification, respectability and authenticity of data. The security is portrayed as the protection so that customers can keep up control over their fragile data. This zone deals with most issues of security and assurance in the particular Internet-based preparing regions. A bit of the standard difficulties for the use of the IoT [11,12] consolidate security issues. Those troubles will be generally in perspective of information security organization structures and what's more on legal foundations. While thinking about the legal arrangement of security and assurance of the IoT [13,14], it must be settled which model of control should be associated. It has been inspected that the possibility of IoT [15,16] relies upon essentially three layers or levels as showed up in the written work. Discusses security for each level and for the improvement of data among levels and besides for each device or structure related constitute the subject of various papers [17–19] for couple of years earlier. Additionally, is said that security must torment the entire model remembering the true objective to stand up to assorted sort of perils. Among these sorts, one can find unverifiable Web/Cloud/Mobile Interface, insufficient approval/approval, unstable system administrations [20,21], nonattendance of transport encryption, assurance concerns, deficient security configurability, problematic programming/firmware and poor physical security. In IoT [22] and [23] and to the extent customer security, the issues fuse the going with: (a) control of individual data, (b) change of insurance advances and the noteworthy headings, (c) measures techniques and programming activities to manage the customers and articles identity. To the extent secrecy, a bit of the issues consolidate the going with: (a) the prerequisite for an easy to use exchange of fundamental, anchored and private information and (b) classification must be an organized into the IoT [24,25] setup process. Heaps of meta-data or impermanent data required for the execution of administrations might be made due to the supervise ment and treatment of tremendous measures of data to guarantee significant information/organization, privacy and genuineness of data. These meta-data or transitory data could have the going with traits: gauge, heterogeneity, et cetera. Application systems for IoT [26,27] must give profitable continuous treatment of data when asked for (customer requests).

5.2.1 WoT-Supported Safe Implementations in Wireless Systems

Technologies such as Internet of Things(IoT) [28,29] enable little devices to offer web-based services in an open and dynamic systems administration conditions on a gigantic scale. End clients or administration shoppers confront a hard choice over which administration to pick among the accessible ones, as security holds a key in the basic leadership process. A base phonetic assessment set is composed [30], in light of which the various fluffy term sets that utilized for portraying security traits are formally dressed and incorporated for figuring a general security estimation of the services. The advance in the zone of installed frameworks has favoured the rise of purported "Smart things "or "Things." These ones consolidate, in a setting of low vitality utilization, different remote correspondence capacities joined with a small-scale controller driving sensors as well as actuators. Mobile phones, associated TVs, smart watches, and so on, are solid cases of keen articles having a place with our regular daily existence. The Internet of Things (IoT) [31,32] conceptualizes this new condition in view of conventional systems associated with objects as particular segments of this present reality [33]. Building, however, a worldwide biological system assembling the diverse IoT conditions, where Things can convey consistently, is a troublesome errand. Since each IoT [34,35] stage utilizes its own heap of correspondence conventions, they normally are not ready to work over the numerous accessible systems administration interfaces, which makes storehouses of clients and Things [36,37]. Web of Things (WoT) has the desire to give a solitary general application layer convention empowering the different Things to speak with each other flawlessly by utilizing the benchmarks and the APIs of the web as an all-inclusive stage. The verbalization among items and the Internet in the event that it speaks to a solid purpose of the WoT, leads likewise this one to acquire every one of the issues of security and protection officially introduced in the Internet. These issues rest with more grounded keenness in this new condition, as a result of its specific qualities. It is along these lines imperative to investigate the manner by which conventional security and protection necessities can be declined in this new condition. In this section, we will endeavor to give a worldwide outline of the currently proposed structures for securing the WoT. This outline covers an investigation of the distinctive dangers and vulnerabilities that an IoT [1,38], in the long run a WoT, design can be presented to. It covers additionally the arrangements proposed to illuminate the problematics identified with the character administration, information secrecy, the approval and the entrance control in a WoT framework.

Late progressions in the Internet, web and correspondence advancements cut crosswise over numerous zones of current living and empowered interconnection of each physical protest, including, sensors and actuators. Web-empowered keen articles enable inventive services and applications for various spaces and enhance usage of assets. Research article [39] proposes an interoperable Internet-of-Things (IoTs) [40] and [41] stage for a brilliant home framework utilizing a Web-of-Objects and cloud engineering. The proposed stage controls the home machines from anyplace and furthermore gives the homes' information in the cloud for different specialist co-ops' applications and investigation.

5.2.2 Semantic Web of Things (WoT)

Urban situations are critically required to wind up more quick witted. In any case, building propelled applications on the Internet of Things requires consistent interoperability. In explore article [42] a water learning administration stage has been proposed which expands the Internet of Things towards a Semantic Web of Things, by utilizing the semantic web to address the heterogeneity of web assets. Evidence of idea is exhibited through a choice help device which use both the information driven and learning based programming interfaces of the stage. The arrangement is grounded in an exhaustive philosophy and run base created with industry specialists. This is instantiated from GIS, sensor, and EPANET information for a Welsh pilot. The web benefit gives discoverability, setting, and importance for the sensor readings put away in an adaptable database. An interface shows sensor information and blame derivation warnings, utilizing the reciprocal idea of serving reasonable lower and higher-arrange learning.

Security episodes, for example, directed dispersed disavowal of administration (DDoS) assaults on control matrices and hacking of manufacturing plant modern control frameworks (ICS) are on the expansion. Article [43] unloads where rising security dangers lie for the mechanical internet of things, drawing on both specialized and administrative viewpoints. Legitimate changes are being introduced the European Union (EU) Network and Information Security (NIS) Directive 2016 and the General Data Protection Regulation 2016 (GDPR) (both to be implemented from May 2018). There is a contextual investigation of the rising brilliant vitality store network to outline, investigate and combine the broadness of security worries at play, and the administrative reactions. Mechanical IoT conveys four security concerns [43] to the fore, for example, valuing the move from disconnected to online framework; overseeing fleeting measurements of security; tending to the execution hole for best practice; and drawing in with infrastructural unpredictability.

The most critical existing issues of security and protection of the Internet of Things (IoT) and Cloud of Things (CoT) [44] ideas particularly secrecy issue are described [4]. With the advancement of pervasive figuring, everything is associated all over, accordingly these ideas have been generally contemplated in the writing. Be that as it may, interruptions and vulnerabilities will be more repetitive because of the frameworks many-sided quality and the trouble to control each entrance endeavor. To handle this issue, scientists have been focussing on different methodologies upholding security and protection. In [4] hazard variables and arrangements with respect to these advances are looked into then present and future patterns are talked about. Table 5.1 shows

TABLE 5.1

Description of WoT Related Security Issues in Diversified Applications

Sl. No.	Literatures	Years	Area of Focus
1	Bo Zhou et al. [30]	2018	Performance evaluation of web service security in IoT and WoT
2	Saad EI J et al. [33]	2017	Security management of web of things Issues
3	Asif Iqbal et al. [39]	2018	Inter operable IoT Framework for smart home using web of objects (WoO) and cloud services
4	Shaun Howell et al. [42]	2018	Decision support system for water utilization in Web of Things Context
5	Urquhart et al. [43]	2018	Industrial web of things and insecurity issues
6	Syrine Sahmim et al. [4]	2017	Security aspects in IoT based Cloud Computing Systems
7	Fang et al. [45]	2017	Searching Issues related to big data using big search approach
8	Xiaofeng et al. [46]	2015	Software Defined Network based WoT Architecture Design and issues
9	Liu et al. [47]	2016	Model design for web use mining in large scale WoT using a graph called request dependency graph approach
10	Elmisery et al. [48]	2018	Deep Learning Mashup in Environmental IoT using Cognitive Privacy
11	Sahzad et al. [49]	2017	Authentication and authorization approaches for IoT Applications
12	Alrawais et al. [50]	2017	IoT in Fog Computing with challenges in security and privacy
13	Pinto et al. [51]	2017	Industrial IoT EDGE devices with design of IIOTEED, a trusted Environment
14	Raggett et al. [52]	2015	Challenges and Opportunities in WoT with review and survey
15	Heuer et al. [53]	2015	Application of web technology in physical world using WoT

some recent Web of Things related research work and mechanism based on security issues which are carried out by eminent researchers across the world.

With the quick improvement of huge information investigation, portable registering, Internet of Things, distributed computing, and person to person communication, the internet has extended to a cross-combined and omnipresent space made up of people, things, and data. Internet applications have developed from Web 1.0 to Web 2.0 and Web 3.0, and web data has seen a hazardous development, which is emphatically advancing the appearance of a worldwide time of enormous information. In this universal the internet, conventional web indexes can never again completely fulfill the developing needs of different kinds of clients. Along these lines, web indexes must make totally imaginative, progressive changes for the up and coming age of hunt, which is alluded to as "large pursuit." Looking into the enormous growth of the web is proving the importance of this work [45].

WoT (Web of Things) coordinates keen devices into Web by reusing and expanding Web norms. While Web innovation makes the designers' activity less demanding, it faces security, administration and proficiency challenges. Article [46] proposes WoT/SDN, the design of asset situated WoT based on SDN (Software Defined Network), in which applications could be produced through asset membership and Mashup with the programmability gave by SDN. The key parts are composed, including Security and Management Controller (SMC), different nuclear services and asset membership language structure. Three applications covering device administration, information access and security assurance are illustrated.

In the Web of Things (WoT) condition, Web movement logs contain profitable data of how individuals associate with keen devices and Web servers. Mining the abundance of data accessible in the Web get to logs has hypothetical and down to earth noteworthiness for some, imperative applications like system improvement and security administration. The primary basic advance of the mining undertaking is displaying the connections among HyperText Transfer Protocol (HTTP) asks for getting to Web articles to examine the conduct of Web customers. Article [47] presents the request dependency graph (RDG), a chart portrayal of the connections among HTTP asks. This article outlines and executes a calculation for essential solicitations recognizable proof, which is a basic errand of Web use mining, in light of the RDG.

5.2.3 Private Data Exchange and Cognitive Platform

Information mashup is a Web innovation that consolidates data from various sources into a solitary Web application. Mashup applications bolster new services, for example, ecological checking. The distinctive associations use information mashup services to consolidate informational indexes from the diverse Internet of Multimedia Things (IoMT) setting based services keeping in mind the end goal to use the execution of their information investigation. Be that as it may, mashup, distinctive informational collections from numerous sources, is a security risk as it may uncover nationals particular practices in various districts. In [48], the creators exhibit a subjective based middleware for cognitive-based middleware for private data mashup (CMPM) to serve a brought together natural observing administration. The proposed middleware is outfitted with disguise systems to protect the security of the blended informational collections from numerous IoMT systems engaged with the mashup application. Moreover, it shows an IoT-empowered information [54] mashup benefit, where the mixed media information are gathered from the different IoMT stages, and after that sustained into a natural profound learning administration keeping in mind the end goal to identify fascinating examples in dangerous zones [55]. The feasible highlights inside every area were removed utilizing a multiresolution wavelet change, and afterward bolstered into a discriminative classifier to extricate different examples. We likewise give a situation to IoMT-empowered information mashup administration and experimentation comes about.

With the beginning of the Internet of Things, little however shrewd devices have turned out to be omnipresent. Despite the fact that these devices convey a considerable measure of register control and empower a few intriguing applications, they need traditional interfaces, for example, consoles, mice, and touchscreens. Accordingly, such devices can't verify and approve clients in recognizable ways. Moreover, not at all like for customary settings, a one-time validation toward the beginning of a session more often than not isn't suitable for the IoT, on the grounds that the application situations are dynamic and a client won't not hold physical control or even consciousness of IoT devices very as promptly as with conventional PCs. Subsequently, clients should be persistently validated and approved. Luckily, the IoT offers fascinating potential answers for meeting these necessities. Article [49] talks about a few difficulties and openings in creating persistent validation and approval approaches for the IoT while additionally exhibiting a contextual analysis of a Wi-Fi-based human confirmation framework called WiFiU.

The intrinsic qualities of Internet of Things (IoT) devices, for example, constrained capacity and computational power [56,57], require another stage to productively process information. The idea of haze figuring has been acquainted as an innovation with overcome any issues between remote server farms and IoT devices. Haze registering empowers an extensive variety of advantages, including upgraded security, diminished data transmission, and decreased dormancy. These advantages make the mist a fitting worldview for some, IoT services in different applications [58,59], for example, associated vehicles and brilliant matrices. By and by, haze devices (situated at the edge of the Internet) clearly confront numerous security and protection dangers, much the same as those looked by conventional server farms. In [50] the creators examine the security and protection issues in IoT situations and propose a system that utilizes haze to enhance the dispersion of authentication disavowal data among IoT devices for security upgrade. They additionally introduce potential research bearings went for utilizing mist registering to upgrade the security and protection issues in IoT conditions.

5.2.4 Protection Method in IoT and WoT Devices

With the coming of the Internet of Things (IoT), security has risen as a noteworthy plan objective for keen associated devices. This blast in availability made a bigger assault surface territory. Programming based methodologies have been connected for security purposes; in any case, these techniques must be stretched out with security-situated advances that advance equipment as the foundation of trust. In article [51], the writers show why TrustZone is turning into a reference innovation for securing IoT edge devices, and how improved TEEs can help meet modern IoT applications ongoing prerequisites. As of not long ago, most endeavors in the field have concentrated on devices basically sensors and actuators-and the correspondence advancements used to get to them. Gartner anticipates that IoT merchants will top US$309 billion in coordinate income by 2020, with a large portion of that cash getting from services (https://www.gartner.com/newsroom/id/2636073). Include Web advancements in with the general mishmash, and it is obviously time to consider how to extend the IoT past item [60,61] storehouses into Web-scale open biological communities in view of open benchmarks, including those for recognizable proof, disclosure, and administration interoperability crosswise over stages from various merchants. As billions of devices interface with the Internet, another Web of Things is developing, with virtual portrayals of physical or conceptual substances progressively open through Web advancements. Accomplishing another period of exponential development, practically identical to the soonest days of the Web, will require open markets, open benchmarks, and the vision to envision the potential for this growing WoT [52]. Similarly as Internet availability has empowered natural data sharing and association through the Web, so the Internet of Things are the reason for the Web of Things [53], empowering similarly straightforward cooperation among devices, frameworks, clients, and applications.

5.3 Confront and Disquiet in IoT Security

Improvement of actually savvy homes, sharp urban regions, and keen devices of omnipresent figuring has risen the Internet of Things (IoT) and rose it as a space of endless effect, potential, and progression, with Cisco

Inc. envisioning to have 50 billion related devices by 2020. Regardless, an immense piece of these IoT devices are certainly not hard to hack and trade off. Much of the time, these IoT devices are constrained in process, storing up, and organize limit, and thusly they are more powerless against ambushes than other endpoint devices, for example, moved cell phones, tablets, or PCs. Research article [53] performs study and sorts clearly comprehended security issues with respect to the IoT layered outlining, in spite of the way that customs utilized for systems administration, correspondence, and association.

In later past, the Internet of Things has been a annoyed as for applications and research. Eventually, there are a phenomenal gathering of hubs related with each other to impact arranged applications in zones, going from diversion to business, to cover alia. These applications trade off our private data about our records, flourishing, and area. This effects us to take thriving measures to accomplish a guaranteed correspondence, where the piece attempt of a message by a noxious client can't trade off our security. This security disguises a particularly wide imagined that can be tended to in various ways. Research [25] spins around the frameworks and cryptographic figurings that can be utilized as a bit of the messages traded among the hubs to make anchor Internet of Things arranges in an approach to manage ensure our correspondences. In this article, we have utilized the Midgar stage to overview the specific potential after effects of standard security strategy identified with cryptography with the clarification behind testing the contrasting blends to discover a reaction for the Internet of Things when it utilizes insecure customs. Examining the outcomes to pick the best arrangement, concerning expenses and security, we derived that the use the RSA, AES and SHA-3 calculations are a true blue trustworthiness to ensure message protection among unbelievable things. This blend offers the smallest consumption– security relationship among every single one of the mixes that we have endeavored in our evaluation. Table 5.2 demonstrates the Issues in IoT Security and their Solution Mechanism in points of interest.

With the broadening scaling down of cell phones, PCs, and sensors in the Internet of Things (IoT) point of view, propping the security and expecting ransomware strikes have wound up being key concerns. Customary security parts are never again critical in light of the joining of preferred standpoint obliged devices, which require all the more figuring power and assets. Ransomware ambushes and security worries in IoT has been appeared in a word [26]. This article at first takes a gander at the ascending of ransomware attacks and layout the related inconveniences. A coherent gathering has been realized by asking for and engineering the writing in context of essential parameters, for example, dangers, necessities, IEEE principles, affiliation level, and progressions and so forth. In addition, a few possible important examinations are spread out to alarm individuals concerning how truly IoT devices are vulnerable against dangers.

TABLE 5.2

Issues in IoT Security and Their Solution Mechanism

Sl. No.	Literatures	Years	Description	Solution and Approach
1	Ahmad khan et al. [53]	2017	Review on IoT security and challenges	Classification of major security issues in IoT
2	Arias et al. [25]	2017	Smart objects security using heterogeneous device platform for IoT	Uses Cryptographic algorithm RSA, AES, SHA-3 for IoT security during message exchange between IoT devices
3	Yaqoob et al. [26]	2017	Emerging security challenges and rise of ramsomware	Finds cause, effect and attack prevention of ramsomware
4	Till et al. [27]	2018	IoT and security concern in Technology	Improvement of security operations in IoT based electronic systems including home appliances.
5	Cerullo et al. [28]	2018	IoT and wireless security in intelligence data centric systems	Surveys on WSN security in smart home associated IoT systems
6	Romdhani et al. [29]	2017	Currently available security scheme for IoT with study and analysis	IoT security is deeply analyzed with kerckhoff's principle of cryptography and use of cryptographic algorithm
7	Scarfo et al. [31]	2018	Cyber security challenges and their relation with IoT	Vulnerability in IoT devices are seriously analyzed and focuses on current trends in cyber security
8	Li et al. [32]	2017	Security importance in IoT Architecture	Significant four layer security architecture of IoT has been described with sensing layer, network layer, service layer and application interface layer
9	Stergiou et al. [23]	2018	An integrated approach of security in IoT and Cloud Computing	Common security concerns related to cloud computing and IoT are discovered to access benefits of their integration
10	Alaba et al. [34]	2017	Detailed analysis on IoT security	Classification of security threats in context of application, architecture and communication is presented

Internet of Things (IoT) is a transformational technology for electronic security frameworks. In different regards, business and private security things were pioneers to IoT and keep sharing different essential qualities of the class. Everything considered, the fast drop in cost for IoT devices and the billions of new IoT devices that are relied on to be displayed in the going with 5–10 years make it a vitality to be battled with to the degree how we consider electronic security structures and the market everything considered. IoT also updates the business endeavors in light of the way that the for all intents and purposes every once in a while alluded to client profit of IoT in the house is for security applications [27]. Everything considered, the move of IoT increment a motivating force for security purchasers in light of the fact that the cash related parts of the IoT business will apply slipping cost weight on security areas, even as they wind up being more able and give more highlights. The wide-scale dissipating of the Internet has been the focal reason for this making configuration, to be specific the use of such general correspondence foundation for empowering machines and amazing articles to give, encourage, and take choices on bona fide word conditions. A study [19] desires the essential technique to deal with the IoT world. Distinctive dreams of this novel viewpoint and device objectives are spoken to and connecting with advancements audited. IoT Security, Safety, and Privacy dangers are familiar and broke down with give an expansive perspective of current issues in light of the decision of this technology. Specific idea is on the Wireless Sensor Network, which tends to the most utilized sensors organize in different zones, for example, Smart Home, giving an overview about its rule particular difficulties, attacks, and related countermeasures.

The fundamental considerations attracted with the security of the Internet of Things (IoT) intertwine arrange; uprightness and realness, receptiveness, no renouncement, and access control, which besides joins perceiving confirmation, attestation, and support [29]. The separations among security and protection are depicted, and furthermore Kerckhoff's oversee of cryptography, and the use of cryptographic calculations. A discussion of key association customs takes in the wake of, considering the ways to deal with oversee association and moreover criteria to think about it, including scrambling, endorsement, extensibility, quality, flexibility, and conspiracy versatility. Nowadays, the headway of the universe of the Internet of Thing is promising the impact of different devices related with the Internet. As demonstrated by Cisco specialists [31] in 2015 there were more than 25 billion devices related and with a projection of more than 50 billion devices related by 2020. Similarly, the new business benchmarks that the Internet of Things advances engage are conveying a super-snappy addition of machine-to-machine correspondences. This is a certified market jump forward moment that opens up a ton open entryways for attempts and, all things considered talking, for the whole society. Characteristically, it augments radically the security issues, which could astound a sizeable bit of Internet of Things' potential points of interest that McKinsey valuates are high. Without a doubt, a present review by HP reports that the 70% of devices contain vulnerabilities. The purpose of this zone is to give a layout of current examples about advanced security concerns and a glance at what the possible destiny of the Internet of Things will report. The four-layer security plan of the Internet of Things (IoT) has been depicted [31]. This building involves the identifying layer, which is fused with IoT end-hubs to recognize and secure the information of related devices; the system layer, which supports the relationship between remote or wired devices; the organization layer, which gives and regulates administrations required by customers and additionally applications; and the application–interface layer, which includes the methodologies for association between the customers and applications. Factors that are basic to the arrangement of the IoT framework are immediately had a tendency to including specific parts, security protection, and business issues. A point by point investigation of IoT and Cloud Computing [34] and [62] with a consideration on the security issues of the two advances can give a responsibility of both. The Internet of things (IoT) has starting late changed into an essential research subject since it encourages unmistakable sensors and articles to discuss especially with each other without human intercession. The necessities for the tremendous scale sending of the IoT are quickly stretching out with a basic security concern. Point by point examination [35] spins around the front line IoT security hazards and vulnerabilities by planning a sweeping review of existing works in the zone of IoT security. The legitimate portrayal of the present security dangers as for application, plan, and correspondence is displayed.

5.4 Conclusion

The two advances Web of Things and Internet of Things are essential specific fragments in different working issues of web related application. Also as Internet organize has engaged natural information sharing and association through the Web, so it outlines the explanation behind the Web of Things, enabling correspondingly essential relationship among gadgets, structures, customers, and applications. Both IoT and WoT pass on knowledge to basic points like transportation, industry, installments, prosperity and different others. The relationship between embedded gadgets and Cloud based web administrations is a conventional circumstance of IoT sending and WoT execution. From the security perspective, in both the two customers and astute gadgets must develop a protected correspondence channel and have a sort of electronic identity in both IoT and WoT technology. To layout this exploration paper completes a whole multidimensional survey on various testing issues in IoT and WoT condition and how the security is executes in brilliant gadgets in these two systems using specific straightforward strategies with an objective that this will include fundamental, fitting and extended zone of research in this amazing region of PC security.

References

1. F. A. Alaba, M. Othman, I. A. Hashem, and F. Alotaibi, Internet of things security: A survey, *Journal of Network and Computer Applications*, 88, 2017, 10–28.
2. M. M. Mahmoud, J. J. Rodrigues, K. Saleem, J. Al-Muhtadi, N. Kumar, V. Korotaev, Towards energy-aware fog-enabled cloud of things for healthcare, *Computers & Electrical Engineering*, 67, 2018, 58–69. doi:10.1016/j.compeleceng.2018.02.047.
3. F. Callegati, S. Giallorenzo, A. Melis, and M. Prandini, Cloud-of-Things meets Mobility-as-a-Service: An insider threat perspective, *Computers & Security*, 74, 2018, 277–295. doi:10.1016/j.cose.2017.10.006.
4. S. Sahmim and H. Gharsellaoui, Privacy and Security in Internet-based Computing: Cloud Computing, Internet of Things, Cloud of Things: A review, *Procedia Computer Science*, 112, 2017, 1516–1522. doi:10.1016/j.procs.2017.08.050.
5. J. Sun, S. Sun, K. Li, D. Liao, A. K. Sangaiah, and V. Chang, Efficient algorithm for traffic engineering in Cloud-of-Things and edge computing, *Computers & Electrical Engineering*, 2018. doi:10.1016/j.compeleceng.2018.02.016.
6. L. Wu, B. Chen, K.-K. Choo, and D. He, Efficient and secure searchable encryption protocol for cloud-based Internet of Things, *Journal of Parallel and Distributed Computing*, 111, 2018, 152–161. doi:10.1016/j.jpdc. 2017.08.007.
7. A. Farahzadi, P. Shams, J. Rezazadeh, R. Farahbakhsh, Middleware technologies for cloud of things-a survey, *Digital Communications and Networks*, 2017. doi:10.1016/j.dcan.2017.04.005.
8. Y. Nan, W. Li, W. Bao, F. C. Delicato, P. F. Pires, and A. Y. Zomaya, A dynamic tradeoff data processing framework for delay-sensitive applications in Cloud of Things systems, *Journal of Parallel and Distributed Computing*, 112, 2018, 53–66. doi:10.1016/j.jpdc.2017.09.009.
9. J. Yang, C. Wang, Q. Zhao, B. Jiang, Z. Lv, and A. K. Sangaiah, Marine surveying and mapping system based on Cloud Computing and Internet of Things, *Future Generation Computer Systems*, 85, 2018, 39–50. doi:10.1016/j.future.2018.02.032.
10. A. R. Al-Ali, I. A. Zualkernan, M. Rashid, R. Gupta, and M. Alikarar, A smart home energy management system using IoT and big data analytics approach, *IEEE Transactions on Consumer Electronics*, 63(4), 2017, 426–434.
11. J. Numms, Salesforce turns to Amazon Web Services for Internet of Things Cloud. https://www.cbronline.com/emerging-technology/. [Last accessed on April 25, 2018].
12. E. White, The Feature and security in the cloud. http://hp.sys-con.com/node/3250225/Cloud and Internet of Things By @HP | @CloudExpo. [Last accessed on April 25, 2018].
13. B. Afzal, M. Umair, G. A. Shah, and E. Ahmed, Enabling IoT platforms for social IoT applications: Vision, feature mapping, and challenges, *Future Generation Computer Systems*, 2017.
14. S. Mukherjee and G. P. Biswas, Networking for IoT and applications using existing communication technology, *Egyptian Informatics Journal*, 19, 2017, 107–127. doi:10.1016/j.eij.2017.11.002.
15. A. Khanna and R. Anand, IoT based smart parking system, *2016 International Conference on Internet of Things and Applications (IOTA)*, Pune, Maharashtra, 2016, 266–270.

16. I. Farris, A. Orsino, L. Militano, A. Iera, and G. Araniti, Federated IoT services leveraging 5G technologies at the edge, *Ad Hoc Networks*, 68, 2018, 58–69.

17. M. Rath, B. Pati, and B. K. Pattanayak, Relevance of soft computing techniques in the significant management of wireless sensor networks, In *Soft Computing in Wireless Sensor Networks*, edited by H. Thanh Binh, N. Dey. Chapman and Hall/ CRC, Taylor & Francis Group, New York, 2018, 86–106.

18. M. Rath and B. Pattanayak, Technological improvement in modern health care applications using Internet of Things (IoT) and proposal of novel health care approach, *International Journal of Human Rights in Healthcare*, 2018. doi:10.1108/ IJHRH-01-2018-0007.

19. M. Rtah, Big Data and IoT-Allied Challenges Associated with healthcare applications in smart and automated systems, HYPERLINK "https://www.igi-global. com/journal/international-journal-strategic-information-technology/1163" *International Journal of Strategic Information Technology and Applications*, 9(2), 2018. doi:10.4018/IJSITA.201804010.

20. M. Rath, B. Pati, C. R. Panigrahi, and J. L. Sarkar, QTM: A QoS task monitoring system for mobile Ad hoc networks. In *Recent Findings in Intelligent Computing Techniques*, edited by P. Sa, S. Bakshi, I. Hatzilygeroudis and M. Sahoo, 707, Springer, Singapore, 2019.

21. M. Rath, B. Pati, and B. K. Pattanayak, An overview on social networking: Design, issues, emerging trends, and security, *Social Network Analytics: Computational Research Methods and Techniques*, Academic Press, Elsevier, London, UK, 2018, 21–47.

22. R. Silva, J. SaSilva, and F. Boavida, A symbiotic resources sharing IoT platform in the smart cities context, *Intelligent Sensors, Sensor Networks and Information Processing (ISSNIP), 2015 IEEE Tenth International Conference*, 2015.

23. C. Chibelushi1, A. Eardley, and A. Arabo, Identity management in the internet of things: The role of MANETs for healthcare applications, *Computer Science and Information Technology*, 1(2), 2013, 73–81.

24. E. Sun, X. Zhang, and Z. Li, The internet of things (IOT) and cloud computing (CC) based tailings dam monitoring and pre-alarm system in mines, *Safety Science*, 50, 2012, 811–815.

25. M. A. Khan and K. Salah, IoT security: Review, blockchain solutions, and open challenges, *Future Generation Computer Systems*, 2017. doi:10.1016/ j.future.2017.11.022.

26. G. Sánchez-Arias, C. G. García, and B. C. G-Bustelo, Midgar: Study of communications security among Smart Objects using a platform of heterogeneous devices for the Internet of Things, *Future Generation Computer Systems*, 74, 2017, 444–466.

27. I. Yaqoob, E. Ahmed, M. H. ur Rehman, A. I. Ahmed, M. A. Al-garadi, M. Imran, and M. Guizani, The rise of ransomware and emerging security challenges in the Internet of Things, *Computer Networks*, 129, 2017, 444–458.

28. S. Van Till, Chapter 8 – Why IoT Matters in Security, In The Five Technological Forces Disrupting Security, *Butterworth-Heinemann*, 2018, 83–95. doi:10.1016/ B978-0-12-805095-8.00008-9.

29. G. Cerullo, G. Mazzeo, G. Papale, B. Ragucci, and L. Sgaglione, Chapter 4 – IoT and Sensor Networks Security, In *Security and Resilience in Intelligent Data-Centric Systems and Communication Networks*, edited by M. Ficco and F. Palmieri, Academic Press, 2018, 77–101. doi:10.1016/B978-0-12-811373-8.00004-5.

30. B. Zhou, Q. Zhang, Q. Shi, Q. Yang, P. Yang, and Y. Yu, Measuring web service security in the era of Internet of Things, *Computers & Electrical Engineering*, 66, 2018, 305–315. doi:10.1016/j.compeleceng.2017.06.020.

31. I. Romdhani, Chapter 7 – Existing Security Scheme for IoT, In *Securing the Internet of Things*, Syngress, Boston, MA, 2017, 119–130. doi:10.1016/B978-0-12-804458-2.00007-X.

32. A. Scarfò, Chapter 3 – The Cyber Security Challenges in the IoT Era, In *Security and Resilience in Intelligent Data-Centric Systems and Communication Networks*, edited by M. Ficco and F. Palmieri, Elsevier, Amsterdam, the Netherlands, 2018, 53–76.

33. S. El Jaouhari, A. Bouabdallah, and J-M. Bonnin, Chapter 14—Security issues of the web of things, In *Managing the Web of Things*, edited by Q. Z. Sheng, Y. Qin, L. Yao, and B. Benatallah, Morgan Kaufmann, Boston, MA, 2017, 389–424. doi:10.1016/B978-0-12-809764-9.00018-4.

34. S. Li, Chapter 5 – Security Requirements in IoT Architecture, In *Securing the Internet of Things*, Syngress, Boston, MA, 2017, 97–108, doi:10.1016/B978-0-12-804458-2.00005-6.

35. C. Stergiou, K. E. Psannis, B.-G. Kim, and B. Gupta, Secure integration of IoT and cloud computing, *Future Generation Computer Systems*, 78, 2018, 964–975.

36. M. Rath and B. K. Pattanayak, Security protocol with IDS framework using mobile agent in robotic MANET, *International Journal of Information Security and Privacy*, 13(1), 2019, 46–58. doi:10.4018/IJISP.2019010104.

37. M. Rath, B. Pati, and B.K. Pattanayak, Mobile agent-based improved traffic control system in VANET. In: Integrated Intelligent Computing, Communication and Security, edited by A. Krishna, K. Srikantaiah, C. Naveena, vol. 771, Springer, Singapore, 2019.

38. N. Akatyev and J. I. James, Evidence identification in IoT networks based on threat assessment, *Future Generation Computer Systems*, 2017.

39. A. Iqbal, F. Ullah, H. Anwar, K. S. Kwak, M. Imran, W. Jamal, and A. ur Rahman, Interoperable Internet-of-Things platform for smart home system using Web-of-Objects and cloud, *Sustainable Cities and Society*, 38, 2018, 636–646. doi:10.1016/j.scs.2018.01.044.

40. I. Romdhani, Chapter 9 – Confidentiality and Security for IoT Based Healthcare, In *Securing the Internet of Things*, Syngress, Boston, MA, 2017, 133–139. doi:10.1016/B978-0-12-804458-2.00009-3.

41. B.-C. Chifor, I. Bica, V.-V. Patriciu, and F. Pop, A security authorization scheme for smart home Internet of Things devices, *Future Generation Computer Systems*, 2017.

42. S. Howell, Y. Rezgui, and T. Beach, Water utility decision support through the semantic web of things, *Environmental Modelling & Software*, 102, 2018, 94–114. doi:10.1016/j.envsoft.2018.01.006.

43. L. Urquhart and D. McAuley, Avoiding the internet of insecure industrial things, *Computer Law & Security Review*, 2018. doi:10.1016/j.clsr.2017.12.004.

44. A. S. Sohal, R. Sandhu, S. K. Sood, and V. Chang, A cybersecurity framework to identify malicious edge device in fog computing and cloud-of-things environments, *Computers & Security*, 74, 2018, 340–354. doi:10.1016/j.cose. 2017.08.016.

45. B. Fang, Y. Jia, X. Li, A. Li, and X. Wu, Big Search in Cyberspace, *IEEE Transactions on Knowledge and Data Engineering*, 29(9), 2017. 1793–1805. doi:10.1109/TKDE.2017.2699675.

46. Q. Xiaofeng, L. Wenmao, G. Teng, H. Xinxin, W. Xutao, and C. Pengcheng, WoT/SDN: Web of things architecture using SDN, *China Communications*, 12(11), 5015, 1–11.

47. J. Liu, C. Fang, and N. Ansari, Request Dependency Graph: A Model for Web Usage Mining in Large-Scale Web of Things, *IEEE Internet of Things Journal*, 3(4), 2016, 598–608.

48. A. M. Elmisery, M. Sertovic, and B. B. Gupta, Cognitive Privacy Middleware for Deep Learning Mashup in Environmental IoT, *IEEE Access*, 6, 2018, 8029–8041. doi:10.1109/ACCESS.2017.2787422.

49. M. Shahzad and M. P. Singh, Continuous Authentication and Authorization for the Internet of Things, *IEEE Internet Computing*, 21(2), 2017, 86–90. doi:10.1109/MIC.2017.33.

50. A. Alrawais, A. Alhothaily, C. Hu, and X. Cheng, Fog Computing for the Internet of Things: Security and Privacy Issues, *IEEE Internet Computing*, 21(2), 2017, 34–42. doi:10.1109/MIC.2017.37.

51. S. Pinto, T. Gomes, J. Pereira, J. Cabral, and A. Tavares, IIoTEED: An Enhanced, Trusted Execution Environment for Industrial IoT Edge Devices, *IEEE Internet Computing*, 21(1), 2017, 40–47.

52. D. Raggett, The Web of Things: Challenges and Opportunities, *Computer*, 48(5), 2015, 26–32. doi:10.1109/MC.2015.149.

53. J. Heuer, J. Hund, and O. Pfaff, Toward the Web of Things: Applying Web Technologies to the Physical World, *Computer*, 48(5), 2015, 34–42.

54. K.-K. R. Choo, M. Bishop, W. Glisson, and K. Nance, Internet- and cloud-of-things cybersecurity research challenges and advances, *Computers & Security*, 74, 2018, 275–276. doi:10.1016/j.cose.2018.02.008.

55. A. Botta, W. de Donato, V. Persico, and A. Pescapé, Integration of cloud computing and internet of things: A survey, *Future Generation Computer Systems*, 56, 2016, 684–700. doi:10.1016/j.future.2015.09.021.

56. https://www.cstl.com/Symantec/Blue-Coat-Advanced-Web-Cloud-Security/. [Last Accessed July 24, 2018].

57. D. D. Sanju, A. Subramani, and V. K. Solanki, Smart City: IoT Based Prototype For Parking Monitoring & Parking Management System Commanded by Mobile App, In *Second International Conference on Research in Intelligent and Computing in Engineering*.

58. R. Dhall and V. K. Solanki, An IoT Based Predictive Connected Car Maintenance Approach, *International Journal of Interactive Multimedia and Artificial Intelligence*.

59. V. K. Solanki, M. Venkatesan, and S. Katiyar, Conceptual Model for Smart Cities for Irrigation and Highway Lamps using IoT, *International Journal of Interactive Multimedia and Artificial Intelligence*.

60. V. K. Solanki, V. Muthusam, and S. Katiyar, Think Home: A Smart Home as Digital Ecosystem, *Circuits and Systems*, 10 (7).

61. V. K. Solanki, S. Katiyar, V. B. Semwal, P. Dewan, M.Venkatesan, and N. Dey, *Advance Automated module for smart and secure City in ICISP-15*, organised by G. H.Raisoni College of Engineering & Information Technology, Nagpur, Maharashtra, on December 11–12, 2015, Procedia Computer Science, Elsevier.

62. Z. Lv, H. Song, P. Basanta-Val, A. Steed, and M. Jo, Next-Generation Big Data Analytics: State of the Art, Challenges, and Future Research Topics, *IEEE Transactions on Industrial Informatics*, 13(4), 2017, 1891–1899.

6

Programming Language and Big Data Applications

Nikita Bhatt and Amit Ganatra

CONTENTS

6.1 Evolution of Big Data

Due to enhancement of technology, a lot of data is being generated [1–3]. Figure 6.1 shows sources of the big data. Initially we had landline phones; now we have smart phones that are making our lives smarter. Apart from that, we used bulky desktop computers for processing a small amount of data (we used floppy and hard disks to store data but now we can store data on the cloud). The generated data is not in the format that can be handled by a relational database [3–5]. Apart from that, the volume of data is also increased exponentially. The Internet of Things (IoT) is another reason for the evolution of big data. For example, the self-driving car is an example of the IoT, which connects physical devices with the Internet and makes the device smarter, and for that a lot of data is generated from the different devices' sensors. Another source for big data is social media, where data is generated in different format [1,6].

6.1.1 Characteristics of Big Data

1. **Volume**

   In today's world, due to enhancement of technology, data is growing rapidly. The *Forbes* 2015 [7] study said that by 2020, accumulated digital universe of data will grow from 4.4 zetabytes today to around 44 zetabytes, or 44 trillion gigabytes.

2. **Variety**

   Different kinds of data is being generated from various sources with different formats, like structured (e.g., a table where a proper scheme is available), semi-structured (JSON, XML, CSV, TSV, email) or unstructured (log, audio, video, image).

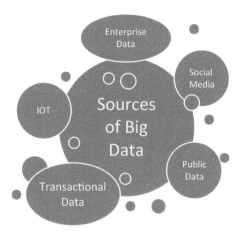

FIGURE 6.1
Different sources of big data.

3. Velocity

Data is being generated at an alarming rate. Velocity is the speed at which the data is created, stored, analyzed, and visualized. In the past, when batch processing was a common practice, it was normal to receive an update from the database every night or even every week. Computers and servers required substantial time to process the data and update the databases. In the big data era, data is created in real time or near real time. With the availability of Internet-connected devices, wireless or wired, machines and devices can pass on their data the moment it is created [4,5,8]. The speed at which data is created currently is almost unimaginable: Every minute we upload 100 hours of video on YouTube. In addition, every minute over 200 million emails are sent, around 20 million photos are viewed and 30,000 uploaded on Flickr, almost 300,000 tweets are sent, and almost 2.5 million queries on Google are performed [2–5,8].

4. Veracity

Veracity refers to the trustworthiness of the data. It is the conformity to facts, with data to test different hypotheses. A lot of data and a big variety of data with fast access are not enough. The data must have quality and produce credible results that enable right action when it comes to end-of-life decision making [9,10].

5. Visualization

Data visualization is representing data in some systematic form, including attributes and variables for the unit of information. Visualization-based data discovery methods allow business users to mash up disparate data sources to create custom analytical views (Figure 6.2).

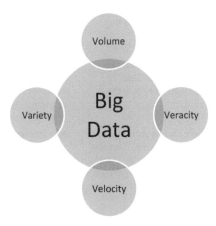

FIGURE 6.2
Characteristics of big data.

6.1.2 Key Big Data Principles

Big data solutions are applicable to structured, semi-structured, and unstructured data from a wide variety of sources, and it becomes ideal when business measures on data are not predetermined. Currently everyone, including academia, industry, and government, are involved in big data projects. It includes all type of data, which requires high-storage and high-performance computing. Big data applies to information that can't be processed or analyzed using traditional processes or tools. Data that is beyond the capacity of the storage and beyond the capacity of the processing power is referred to as big data [1,2,11]. Table 6.1 shows the comparison between traditional data processing and big data analytics.

6.1.3 Research Issues on Big Data

The following are the open research issues on Big Data [12–14].

1. Data Volume (Scale) and Processing

According to *Forbes* 2015 [7] report, data generated in the past is years is more than that of all previous history in total. By 2020, about

TABLE 6.1

Traditional Data Processing versus Big Data Analytics

	Traditional	Big Data Analysis
Memory Access	Data is stored in centralized RAM and can be efficiently scanned several times.	Data can be stored in high distributed data sources. In case of huge, continuous data streams, data is accessed only in single scan.
Computational Processing and architectures	Serial centralized processing: A single-computer platform that scales with better hardware is sufficient.	Parallel and distributed architectures may be necessary. Cluster platforms that scale with several nodes may be necessary.
Data Types	Data source is relatively homogeneous. Data is static and of reasonable size.	Data comes from multiple sources that may be heterogeneous and complex. Data may be dynamic and evolving. Adopting to data changes may be necessary.
Data Management	Data format is simple and fits in a relational database or data warehouse. Data access time in not critical.	Data format is usually diverse and may not fit in a relational database. Data may be greatly interconnected and needs to be integrated from several nodes. Often special data systems are required that manage varied data formats (NoSQL, Hadoop, etc.). Data access time is critical for scalability and speed.

1.7 MB of new information will be created every second for every person. Storing this exponential data into a traditional system is not possible. So a distributed file system is a better way to store such large amounts of data. Storage is one problem, but along with that processing is also one of the crucial operations, as data is available in different formats. Therefore, the system must present that stores a variety of data that is generated from various sources.

2. Security and Privacy

Data security and data privacy are critical issues when migrating big data to a cloud environment. Many technical solutions using data encryption are applied: Data encryption affects the performances of big data processing, thus researchers are still working for further improvements to this issue.

Today services demand us to share private data, but outside their demand one cannot know what is the meaning of sharing data here and how the data that is shared can be linked [15].

3. Data Transmission

Today hard-disk capacity is increasing, but disk transfer capacity or speed is not increasing at the same rate. While transferring large-scale data to the cloud, the capacity of the network bandwidth is the main issue. Over the years many algorithmic proposals and improvements have been applied to minimize cloud upload time; however, this process still remains a major research challenge.

6.1.4 Big Data Applications

Following are the applications where big data is widely used [16–18].

6.1.4.1 Healthcare

Usually, the healthcare organizations like to utilize big data; the reason is because of their partial ability to regulate and merge data. In contrast to that, now analytics have been upgraded in healthcare by giving personalized remedies and narrow analytics. Investigators mining the information to check what treatment measures are simpler for specific conditions determine patterns associated with drugs' multifaceted effects, and these advances differentiate necessary information, which will facilitate patient care and scale back prices [19].

6.1.4.2 Manufacturing

Manufacturing demands a huge volume of information and innovative prediction tools for efficiently converting data into meaningful information. Primary advantages in manufacturing industries are as follows:

- Product quality and defects tracking
- Supply planning
- Manufacturing process defect tracking
- Output forecasting
- Increasing energy efficiency
- Testing and simulation of new manufacturing processes
- Support for mass-customization of manufacturing

6.1.4.3 Media and Entertainment

Some of the applications for the area of the media and entertainment industries are as follows:

- Predicting what the audience wants
- Scheduling optimization
- Increasing acquisition and retention
- Ad targeting
- Content monetization and new product development

6.1.4.4 Internet of Things (IoT)

In IoT, Mappings of device interconnectivity are utilized by numerous corporations and governments to extend potency. IoT is additionally more and more adopted as a way of collecting sensory knowledge, which is employed in medical and producing contexts [19].

6.1.4.5 Government

- **Cyber Security and Intelligence** The government has planned a cyber-security R&D plan that trusts the capability to analyze huge data sets to increase the security of computer networks. The "National Geospatial-Intelligence Agency" is creating a "Map of the World" that combines and analyzes data from different sources like satellites and social media.
- **Crime Prediction and Prevention** Police departments will control advanced, time-period analytics to supply intelligence about unjust practices that may be able to perceive criminal behavior.
- **Pharmaceutical Drug Evaluation** Massive information technologies may cut back analysis and implementation prices for pharmaceutical manufacturers at large [McKinsey report]. The 21st century

National Institutes of Health (NIH) use massive information technologies to get massive knowledge to judge medication.

- **Scientific Research** The NSF (National Science Foundation) has started to do the following:
  - Develop novel methods for mining information from data.
  - Implement novel methods to education.
  - Develop a novel structure to manage, curate, and serve data to communities.

- **Weather Forecasting** The continuous incoming-stream data regarding weather is very large big data that requires predictions to be done in real time. The NOAA (National Oceanic and Atmospheric Administration) works on it and analyzes terabytes of data daily.

- **Tax Compliance** Tax organizations also use big data to analyze heterogeneous information from different sources to recognize suspicious events and actions.

- **Traffic Optimization** Collected traffic information from sensors helps in optimizing traffic. GPS devices and video cameras are useful for the same. Analysis also helps us by altering routes in real time.

6.2 Hadoop Framework: Solution to the Problem

Hadoop is a framework that allows us to store and process large data sets in parallel and distributed fashion [20]. It has basically two parts: (1) HDFS (Hadoop Distributed File System), which allows for the dumping any kind of data across the cluster; and (2) MapReduce, which allows parallel processing of the data stored in HDFS. It is nothing but the processing unit of Hadoop.

6.2.1 HDFS (Hadoop Distributed File System)

A large file is impractical to store on a single machine. So we split the whole file into different partitions, and we store those partitions on different machines that are called workers, and we store multiple copies of the same partition on different workers. In order to access the large file, we can retrieve those different partitions many different ways. There are two advantages of it. One is we can access the multiple partitions in the single file concurrently, so the root speed is actually faster than accessing the single file on a single machine. Second, if any of those workers fail, we still can retrieve a larger file. So HDFS is the backhand file system to store all data, using the MapReduce program.

6.2.2 HDFS Architecture

HDFS follows the master-slave architecture, and it has the following elements.

- **Name node** It is a centralized machine called name node, which is controlling various data nodes which is nothing but the commodity hardware. It has all the information related to the data stored into the Data Node. Figure 6.3 shows the relation between name node and data node. It is nothing but the master server that does the following tasks:
  - Acts as Master daemon
  - Maintains and manages data nodes
  - Records metadata (e.g., location of blocks stored, the size of the file, permissions, hierarchy)
  - Receives heartbeat and block report form all the data nodes
- **Data node** It stores the data that is in the Hadoop cluster. It regularly sends a heartbeat rate to the Name node. It has the following features:
  - Slave daemons
  - Stores actual data
  - Serves read and write requests from the clients
- **Block** Each file is stored on HDFS as blocks. The default size of each block is 128MB in Apache Hadoop.

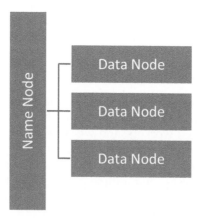

FIGURE 6.3
Name node and data node.

6.2.3 MapReduce

MapReduce is a powerful system that can perform on big data using a distributed environment. Hadoop is the Java implementation of MapReduce. There are many software tools that have been developed to facilitate development effort for data science tasks such as data processing, extraction, and transform and loading process. In a nutshell, Hadoop and MapReduce enable a powerful big data ecosystem by providing the combination of all these things. In fact, MapReduce or Hadoop are big data systems that provide the following capabilities:

- Distributed storage for large datasets through Hadoop Distributed File System
- Distributed computation through programming Interface MapReduce
- Fault-tolerance systems in order to cope with constant system failures on large distributed systems that are built on top of commodity hardware.

MapReduce was proposed by Jeff Dean and Sanjay Ghemawat from Google in 2004 and was implemented inside Google as proprietary software for supporting many of their search engine activities. Later on, Apache Hadoop was developed, which is an open-source software that mimics the original Google's MapReduce systems. It is written in Java for distributed file storage and distributed processing of very large data sets. To program on Hadoop systems, it is necessary to use the programming abstraction MapReduce, which provides a very limited but powerful programming paradigm for parallel processing on a large data set. The entire algorithm running on MapReduce or Hadoop has to be specified as MapReduce programs. The reason for this very constrained programming model is to support super scalable, parallel, and fault-tolerance implementation of data processing algorithms that can run on large data sets. To utilize Hadoop and MapReduce, it is necessary to understand and master common patterns for computation using MapReduce.

The fundamental pattern for writing a data mining or machine learning algorithm using Hadoop is to specify the machine learning algorithms as computing aggregation statistics. So it is necessary to implement machine learning algorithms for identifying the most common risk factors of health failure. And it is also necessary to decompose the algorithm into a set of smaller competition units. For example, we want to extract the list of risk factors related to heart failure that appeared in each patient's record. The result from this map function will be aggregated by a reduce function. For example, instead of listing the risk factors for each patient, we want to compute the frequency of each risk factor over an entire population. Then in that case, the entire reduce function would do that by performing the aggregation statistic on results from the map function. This process is quite abstract.

6.2.4 MapReduce Algorithm

There are three stages in execution of the MapReduce program: map stage, shuffle stage, and reduce stage.

1. **Map Stage**

   The task of map is to process the incoming data. Usually the incoming data is in file or directory form, which will be stored in the Hadoop file system. Then the file is delivered to the mapper function one line at a time. The mapper processes the data and makes several minor pieces of data.

2. **Reduce Stage**

   Reduce is the mixture of the **shuffle** stage and the **reduce** stage. The task of the reduce phase is to process the data that is the outcome of the mapper, and the generated output will be stored in the HDFS [20].

   a. In the task of MapReduce, Hadoop will send the map and reduce job to the suitable servers in the cluster.

   b. The framework copes with the all the small print of data passing resembling provision tasks, confirmatory task completion, and repetition knowledge around the cluster between the nodes.

   c. To cut down the network traffic, the majority of the operation is carried out on nodes with data on native disks (Figure 6.4).

That is mainly how the Hadoop framework is presented. The Hadoop platform is now composed of a number of related projects as well—Pig, Hive, HBase, Spark, Mahout, and more [11]:

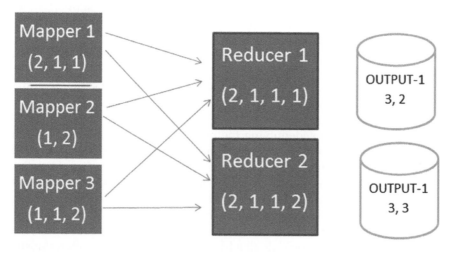

FIGURE 6.4
MapReduce.

- **Apache Hive:** Hive could be a Hadoop-based information processing structure authored by Facebook. It permits users to write down queries in a very SQL-like language called Hive-QL, that are then changed to MapReduce. This enables SQL programmers with no MapReduce expertise to use the warehouse and makes it easy to incorporate with business intelligence and image tools. It gets the Hive SQL string from the client, and the parser phase converts it into parse tree representation. The logical query plan generator converts it into logical query representation and prunes the columns early and pushes the predicates closer to the table. The logical plan is converted to a physical plan and MapReduce jobs.
- **Pig Latin:** Pig Latin is a Hadoop-based language developed by Yahoo. It is relatively easy to learn and is adept at very deep, very long data pipelines (a limitation of SQL). The main goal of Pig Latin is to reduce the time of development. There are certain important operators:
  - FOREACH: To process a transformation over every tuple of the set. It is possible to parallelize this operation, and the transformation of one row should depend on another.
  - COGROUP: It is used to group related tuples of multiple data sets, and it is similar to the first step of join.
  - LOAD: It is used to load the input data and its structure and store to save data in the file.
- **HiveQL:** This is an open-source solution built on Hadoop. The queries look similar to SQL and also have extensions on them. HiveQL stores the metadata in an Relational Database Management System (RDBMS). The meta store acts as the system catalog for Hive. It stores all the information about the tables, their partition, the schema, and so on. Without the system catalog, it is not possible to impose a structure on Hadoop files. Facebook uses MySQL to store this metadata, because information has to be served fast to the compiler.
- **JAQL:** JAQL is a declarative scripting programming language and used over Hadoop's MapReduce framework. It has a simple, flexible data model that handles semi-structured documents but also structured records validated against schema.
- **HBase:** HBase is a traditional database that permits for low-delay, faster lookups in Hadoop. It adds transactional abilities to Hadoop, which permit the user to make updations, insertions, and deletions. Examples of businesses that use HBase are eBay and Facebook.
- **Flume:** Flume is an architecture for calculating Hadoop with data. Proxies are populated through different Web servers and application servers; for example, to gather information and incorporate it into Hadoop.

- **Oozie:** Oozie is a kind of processing system that allows users to provide a sequence of jobs generated in several languages, such as MapReduce, and to afterward intelligently associate with each other. For example, it permits users to provide a specific query to be triggered post completion of preceding jobs.
- **Ambari:** Ambari is a Web-based tools for installing, controlling, and observing Hadoop clusters. The implementation is run by Hortonworks' engineers, who consider Ambari in Hortonworks' data platform.
- **Avro:** Avro is a data series structure that permits converting the logical architecture of Hadoop files. It is practiced at analyzing data and executing removed procedure calls.
- **Mahout:** Mahout is a library that takes the best widespread data-mining algorithms for executing different task like clustering, statistical modeling, and regression and develops them using MapReduce.
- **Sqoop:** Sqoop is a linking tool for transferring data from non-Hadoop data storage like RDBMS and data warehouses into Hadoop. It permits users to provide the target place in Hadoop and train Sqoop to transfer data from an RDBMS to the target.
- **HCatalog:** HCatalog is a non-distributed data dictionary/metadata that controls and shares services for Hadoop. It permits the integrated outlook of all data in Hadoop clusters and permits various tools like Pig and Hive to execute data elements without being required to know the physical location of data in the cluster.
- **BigTop:** BigTop is basically an architecture for wrapping and interoperability testing of Hadoop's subproject. Here the aim is to enhance the Hadoop platform.

6.2.5 Two Ways to Process Big Data

A broad division of processing big data can be seen as batch data processing and real-time data processing. Choosing between which kind of processing to be used is purely based on the nature of data to be handled.

6.2.5.1 Batch Data Processing

This approach of data processing usually involves collection of input data over a period of time. Then this input data is divided into batches and processed. For example, in an ATM, the customer record may be updated only at end of day. In the context of big data technology, there is a need to process huge volumes of large data sets. Data is divided into batches and processed by clusters of processors to get the best result out of the big data. The main idea behind this model of data processing is to achieve high throughput when processing large data sets. This model of processing is extremely useful when the business requirements deal with historic large data sets. For example, big data analytics efficiently uses this model to achieve high throughput.

6.2.5.2 Real-Time Data Processing

This type of processing usually processes the data as it arrives and produces the output instantly. Thus it involves continuous input, process, and output of data at near real time or with very low latency for proper functioning of the system. For example, verification of a PIN number at an ATM is real-time data processing. In the context of big data technology, real-time data processing usually processes the incoming streams of data from various heterogeneous systems. These streams of data input must be processed as they arrives, which is on other end of spectrum when compared with batch data processing in which input data is collected over a period of time and batched for processing. Hence the programming paradigm used for batch data processing can never be used for real-time data processing. For instance, the MapReduce programming model cannot be used for real-time data processing because the map phase will never end due to continual input streams of data, and the reduce phase won't start until the map phase ends its activity.

6.3 Programming Languages

A programming framework is the basic style and interfaces for developers to put in writing executing programs and applications [21–23]. In big data, users concentrate on writing data-driven parallel programing, which may be on a massive scale and in distributed environments. There is a range of programming frameworks to be presented for a large body of knowledge with completely different emphasis and benefits. Massive data processing, widely known as big data," has got advantages over programming languages concepts. As the large data size has grown to require distributed processing, possibly on heterogeneous platforms, there is a demand for efficient programming models and techniques. Ideas from programming languages play an important role in a range of advanced applications of databases, in database system implementation, distributed programming (MapReduce), streaming computation, and high-performance (Graphics Processing Unit (GPU)/multicore) computation. In programming languages for big data analytics, a lot of programming languages are provided that take care of all variant tasks in this field, but they can be divided into two groups [24–27]: high level and low level. These levels are distinguished by the analytical usage of these languages.

6.3.1 High Level

In high-level programming languages, there are many languages for analytical tasks; some of them need a long time to be able to use them, and others are able to be learned within a few weeks. But being professional in any of these languages is a hard task and needs a lot of time (many years).

6.3.1.1 Python

Python language is one of the famous data-analyzing languages that data scientists use to focus on in their research. The high-level interactive nature of this language and its scientific ecosystem libraries make it the preferred choice for developing analytical algorithms and exploring the hidden facts in the data. It is an interpreted programming language, so it is slower than compiled-based language like C++. In any application with very low latency or demanding resource utilization requirements, the time spent programming in a lower-level language like C++ to achieve the maximum possible performance might be time well spent.

Python is an interesting language for constructing concurrent and multi-threaded applications with several CPU-bound threads. The explanation for this is often that it's what's called the "global interpreter lock" (GIL), a mechanism that stops the interpreter from implementing many Python instructions at the same time. Python has many valuable libraries such as the following:

1. **NumPy**

   NumPy is short for "Numerical Python"; it's the base data structure and the fundamental package in Python language. Knowing that all input data in Python is represented as a NumPy array makes it easy to infer that all libraries in this language are built on top of this package. NumPy provides the following features:

   Ndarray: An efficient and fast multidimensional array object.

   a. A long list of functions to manipulate with arrays through element-wise computations with them or by providing mathematical operators between arrays.

   b. Tools for reading and writing array-based data sets to disk.

   c. Fourier transform, linear algebra operations, and random number generation.

   d. Tools to integrate other languages' code (C, C++, and FORTRAN) in Python.

   Apart from quick array-processing abilities that NumPy append to Python, another important uses in analysis is as a container for data to be exchanged between algorithms and libraries. For numeric data, NumPy arrays are most effective for storing and changing data compared to other built-in Python data structures. Apart from that, libraries generated in a language like C or FORTRAN can function on the data sets stored in a NumPy array without copying data into memory.

2. Pandas

Pandas provide high-level data structures and functions designed to make working with structured or tabular data fast, easy, and inexpensive. Pandas blend the high-performance array computing ideas of NumPy with the flexible data manipulation capabilities of spreadsheets and relational databases (such as SQL). It provides sophisticated indexing functionality to make it easy to reshape, slice and dice, perform aggregations, and select subsets of data. Since data manipulation, preparation, and cleaning is such an important skill in data analysis, Pandas is one of the primary libraries that is widely used. This package support the scientists and focuses their tasks on structured data, with its prepared and pre-designed functions and the rich data structures to make these tasks easy and significant.

3. Matplotlib

Matplotlib is for visualization data and drawing expressive plots; this tool is very popular and effective for these tasks, especially for 2D plots.

4. IPython

IPython is an interactive computing and development environment; it's used to maximize your productivity in both interactive computing and software development. It also includes a rich GUI console with inline plotting; a Web-based interactive notebook format; and a lightweight, fast parallel computing engine. It's a consolidation on the Python shell to hasten the writing, testing, and debugging of Python code.

5. SciPy

SciPy is a collection of packages of efficient algorithms for linear algebra, sparse matrix representation, special functions, and basic statistical functions. These packages include the following:

scipy.integrate	scipy.linalg	scipy.optimize	scipy.signal
scipy.sparse	scipy.special	scipy.stats	scipy.weave

6. Cython

Data scientists can use Python syntax and high-level operations, and they increase compiling performance to reach the performance of the compiled languages by using this package because it combines C with Python.

6.3.1.2 R Language

R is a programming language and environment commonly used in statistical computing, data analytics, and scientific research. It is one of the most popular languages used by statisticians, data analysts, researchers, and marketers to retrieve, clean, analyze, visualize, and present data. Due to its expressive syntax and easy-to-use interface, it has grown in popularity in recent years. It is an extremely versatile open-source programming language for statistics and data science. In an R system, you can do any kind of statistical computation by using functional-based syntax or program-based code with very powerful debugging facilities, and this language has many interfaces with other programming languages. The resulting statistics can be displayed by using the high-level graphical tool in R. For statistical data analysis, R is an open-source software platform. Largely because of its open-source nature, R has been speedily adopted by statistics departments in universities around the world, which are attracted to its extensible nature as a platform for academic research. Following are the features of R language:

- A short and slim syntax to accelerate your tasks on your data.
- It has a variant formats for loading and storing data for both local and over-Internet tasks.
- Ability to perform your tasks in memory by using a consistent syntax.
- A long list of tools (functions, packages) for data analysis tasks; some of them are built-in and the rest are open source.
- It has different easy manners of representing the statistical results in graphical methods and the ability to store these graphs on the disk.
- Ability to automate analyze and create new functions and to extend the existing language features.
- Users don't need to reload their data every time, because the system saves the data between the sessions and saves the history of their commands (Table 6.2).

TABLE 6.2

Difference Between Python Language and R Language

Features	Python	R
Libraries	**Numpy,** SciPy., Pandas, Matplotlib, Scikit Learn, etc.	Over 7500 libraries for many domains
Compiler	Interpreted language	Interpreted language
Learning	Suitable to learn	Harder to learn
IDE	PyCharm, Spyder, Anaconda	RStudio, Red-R
Speed	Slow comparing with R	Fast, especially after new computation algorithms

6.3.1.2.1 Real-World Applications of R Programming

1. **Data Science**

   R provides data scientists a power that permits gathering data in real time, executing statistical and different analyses, generating imaginings, and connecting the results to the end user.

2. **Statistical Computing**

   R is a well-known programming language, and it was primarily constructed by statisticians for statisticians. It has a lot of package sources with more than 9000 packages. R's easy-to-use syntax permits researchers, including those who are from fields other than computer science, to speedily ingress, clear, and analyze data from various data sources. R also supports charts to visualize and understand different data sets.

3. **Machine Learning**

   R is useful in machine learning, as it has many packages for machine-learning tasks such as linear and nonlinear regression, decision trees, and linear and nonlinear classification.

6.3.1.3 SAS

SAS helps solve business problems by being the best at applying advanced analytics, whether it's predictive analytics, forecasting, optimization, or a combination of some or all of them, in order to improve business processes and deliver more valuable data-driven information to decision-makers, so that they can make the best decisions possible to help grow their organizations.

SAS is the leading analytical software on the market. The SAS system is divided into two main areas: procedures to perform analysis and fourth-generation language that allows users to manipulate data. This is known as the DATA step. The SAS system processes data in memory when possible and then uses system page files to manage data of any size efficiently. SAS software (with its language) is a famous and common solution for accessing, transforming, and reporting data by using its flexible, extensible, and Web-based interface. The SAS analytics platform consists of many analytical applications that form this application framework and make it a very useful tool for all data scientists in most of their tasks. The main useful analytical applications are the following:

- **SAS Text Miner** is a plug-in that can be added to the SAS Enterprise Miner environment, because it facilitates the main concept of text mining, which is the prediction aspect, and supports it with a very rich set of tools. SAS Text Miner is able to manipulate with different sources of textual data:
  - Normal local text files.
  - Retrieved text from SAS data sets or other external databases.
  - Files on the Web.

6.3.1.4 MS SQL Server

MS SQL Server is a very famous solution for traditional relational databases, and it has very good tools for drawing Entity Relationship Diagrams (ERD) and also for optimizing queries by using graphical tools that explain how MS SQL does this task (query) and the steps of this task. MS SQL Server 2012 packages provide data scientists with full tools for data integration, visualization solutions, a rich business intelligence suite, and the ability to connect with Apache Hadoop and Hive through an efficient connector.

6.3.2 Low Level

Sometimes the high-level data science platform is not enough for a particular analytics task, and data scientists need to go to a lower-level statistics/programming language (low level in analytical tasks). Especially when they find that the code in these high-level languages is almost slower than code written in a compiled language like Java or C++, data scientists need highly concurrent, multi-threaded applications and applications with many CPU-bound threads; they build and create their own solutions, depending on these low-level languages. There are many statistical solutions like SPSS, STATA, MiniTab, Statistica, and more. STATA is very strong solution but it's not easy for nonstatistical users to work with; for that we focus in this part on the most famous solution, which is SPSS because it is easy to use, and even nonstatistical users are able to use it. Table 6.3 shows comparisons among the most important statistical software.

TABLE 6.3

Comparison Between the Most Important Statistical Software

Features	SPSS	STATA	MiniTab	Statistica
Usage	• Easy-to-use by menus • Non-statisticians are able to use it easily.	• Easy-to-use by command or menus. • Some tasks need code. • Needs strong statistical experience.	Needs statistical experience, but it offers some help by Assistant menu.	Needs strong statistical experience.
Learn	Very easy to learn.	Needs longer than SPSS.	Not easy to learn.	Not easy to learn.
Operating System	Windows, Mac OS, Linux.	Windows, Mac OS, Linux.	Windows.	Windows.
Scripting languages	R, Python, SaxBasic	ado, Mata		R, Statistica Visual Basic (SVB)
ANOVA	All kinds	All kinds (one-way, two-way, MANOVA, GLM, mixed model, ost-hoc, Latin squares)	All kinds except mixed model	All kinds except mixed model

References

1. Kumar, B. S., J. M. Kumar, P. S. Rao, "Unified big data lambda architecture with Hadoop/Flume/Spark SQL, Streaming/Scala/Cassandra," *International Journal of Innovative Research in Computer and Communication Engineering*, 4(6), 2016.
2. Kulkarni, M., and Y. H. Lu. "Beyond big data—rethinking programming languages for non-persistent data." *Cloud Computing and Big Data (CCBD), 2015 International Conference on*. IEEE, 2015.
3. Huda, M. et al. "Big data emerging technology: Insights into innovative environment for online learning resources." *International Journal of Emerging Technologies in Learning (iJET)*, 13(1), 2018: 23–36.
4. John Walker, S. "Big data: A revolution that will transform how we live, work, and think," 33, 2014: 181–183.
5. Wu, X. et al. "Data mining with big data." *IEEE Transactions on Knowledge and Data Engineering*, 26(1) 2014: 97–107.
6. Wu, X., X. Zhu, G.-Q. Wu, W. Ding. "Data mining with big data." *IEEE Transactions on Knowledge and Data Engineering*, 26(1), 2014: 97–107.
7. Marr, B., Big data: "20 mind-boggling facts everyone must read." Available at https://www.forbes.com/sites/bernardmarr/2015/09/30/big-data-20-mind-boggling-facts-everyone-must-read/#1a94b88817b1 (September 30, 2015).
8. Ramírez-Gallego, S. et al. "Big Data: Tutorial and guidelines on information and process fusion for analytics algorithms with MapReduce." *Information Fusion* 42, 2018: 51–61.

9. Walkowiak, S. *Big Data Analytics*. Birmingham, UK: R. Packt Publishing, 2016.
10. Kalyvas, J. R., and D. R. Albertson. "A big data primer for executives." *Big Data: a Business and Legal Guide*. Boca Ratón, FL: CRC Press, 2015.
11. Chavan, V., and R. N. Phursule. "Survey paper on big data." *International Journal of Computer Science & Information Technology*, 5(6), 2014: 7932–7939.
12. Singh, B. A., M. S. Lakshmi, S. P. Kumar. "Challenges and outlook with big data." *International Journal of Computer Engineering in Research Trends*, 2(8), 2015.
13. Labrinidis, A., and H. V. Jagadish. "Challenges and opportunities with big data." *Proceedings of the VLDB Endowment*, 5(12), 2012: 2032–2033.
14. Sivarajah, U. et al. "Critical analysis of big data challenges and analytical methods." *Journal of Business Research*, 70 2017: 263–286.
15. Agrawal, D., P. Bernstein, E. Bertino, S. Davidson, U. Dayal, M. Franklin, J. Gehrke et al. "Challenges and opportunities with big data 2011–1." (2011).
16. Lin, D. "Application of a big data platform in the course of Java language programming." *International Journal of Emerging Technologies in Learning (iJET)*, 11(10), 2016: 16–21.
17. Wang, Y., L. A. Kung, and T. A. Byrd. "Big data analytics: Understanding its capabilities and potential benefits for healthcare organizations." *Technological Forecasting and Social Change*, 126, 2018: 3–13.
18. Nita, S. L., and M. Mihailescu. *Practical Concurrent Haskell: With Big Data Applications*. New York: Apress, 2017.
19. Anagnostopoulos, I., S. Zeadally, and E. Exposito. "Handling big data: Research challenges and future directions." *The Journal of Supercomputing*, 72(4), 2016: 1494–1516.
20. Polato, I., R. Ré, A. Goldman, and F. Kon. "A comprehensive view of Hadoop research—A systematic literature review." *Journal of Network and Computer Applications*, 46 2014: 1–25.
21. Siddiqui, T., M. Alkadri, N. A. Khan. "Review of programming languages and tools for big data analytics." *International Journal of Advanced Research in Computer Science*, 8(5), 2017.
22. Cheney, J., and T. Grust. "Special issue on programming languages for big data." editorial. *Journal of Functional Programming*, 28, 2018.

Books

23. Cheney, J., T. Grust, D. Vytiniotis. "Programming languages for big data (PlanBig)(Dagstuhl Seminar 14511)." *Dagstuhl Reports*, 4(12). Schloss Dagstuhl-Leibniz-Zentrum fuer Informatik, 2015.
24. Belcastro, L., F. Marozzo, and D. Talia. "Programming models and systems for big data analysis." *International Journal of Parallel, Emergent and Distributed Systems*, 2018: 1–21.
25. Schmidt, D. et al. "Programming with BIG data in R: Scaling analytics from one to thousands of nodes." *Big Data Research* 8 (2017): 1–11.
26. Yang, C. et al. "Big Data and cloud computing: Innovation opportunities and challenges." *International Journal of Digital Earth* 10(1) 2017: 13–53.
27. Zomaya, A. Y., and S. Sakr, Eds. *Handbook of Big Data Technologies*. Cham, Switzerland: Springer, 2017.

7

Programming Paradigm and the Internet of Things

Sourav Banerjee, Chinmay Chakraborty, and Sudipta Paul

CONTENTS

7.1 IoT Programming Tools

7.1.1 Introduction

This chapter entails a comparative discussion of the programming tools [1] behind the built of IoT. This discussion will be solely dependent on the approaches and problems discussed in the introduction section. IoT itself is in its infant age; therefore, many frameworks are new or in their building or experimental stages. Therefore the main focus in this section will shed light on the knowledge of the respective design, implementation, and validation planning that gives the opportunity for enhancing software legibility, extensibility, and portability. The discussion will eventually lead to domain complexity and the trouble with maturity. These are the biggest challenges in developing programming frameworks. Therefore a list of essential features of programming framework [2] of IoT can be given as follows:

1. *Fault tolerance*

   The most significant feature of a highly fault-tolerant system is that it might continue at the same level of performance even though one or more components or devices have failed or are not responding. Therefore the basic characteristics of fault tolerance require (a) fault isolation to the failing component, (b) fault containment to prevent propagation of the failure, and (c) availability of reversion modes. It is expected to have a frequent system partitioning and loss of connection with the variation of device mobility across a vast geographical area under IoT. Therefore the programming tool should permit programmers to generate applications that can heal their connections and in the time of partition of the network, the traveling between online and offline states become as light-footed as can be.

2. *Lightweight footprint*

   The footprint of an embedded system is the total size of the system. Therefore, two main constraints that will ensure lightweight programming frameworks are (a) lower runtime overhead and (b) needed effort by the programmers.

3. *Maintain the latency-sensitivity*

   The meaning of latency is "the delay before a transfer of data begins following an instruction for its transfer." IoT has well-distributed applications and devices over a vast geographically distributed area, which has definitely led to latency-sensitivity. Therefore squeezing all the computations to the cloud will accelerate the latency more and more. These requirements can only be supported effectively with the adequate runtime along with a proper programming tool. The other essential features are *coordination, heterogeneity,* and *scalability.*

7.1.2 Overview of Programming Approaches of the Frameworks

1. **Node-centric approach of programming:**

   This labor-intensive and non-portable approach expects that the application developer will program each application development, interlinked between nodes, sensor data collection and analyzing, and providing commands to the respective nodes of the actuator. It assures more control over the programming.

2. **Model-driven approach:**

   This approach is used to provide horizontal and vertical separation from a distributed database point of view. To reduce the application development complexity, vertical separation is needed, whereas horizontal separation minimizes the complexity of the system for the different system views.

3. **Database approach:**

   Every node of the system is thought of as a part of a virtual database. The developer can issue the queries on sensor nodes. This approach does not provide the application-logic, which is of minor use in IoT application deployment.

4. **Macro programming:**

   Macro programming is used to specify high-level communication; abstractions are provided and application logic is also specified. Microprogramming ensures the hiding of minimum-level facts that will assist the fast progress of different applications.

7.1.3 Examples of IoT-Programming Tools

7.1.3.1 Mobile Fog

Fog computing is a new model proposed by Cisco. Here, the inclusive logic for applications is executed roundly on the whole network, guaranteeing the invoking of the dedicated nodes and the routers [1]. Based on the geographical nearness and hierarchy, an application will contain distributed processes in every part of the computing framework, cloud, and edge devices respectively. Each process in the application performs sensing, actuation, and aggregation according to its geographical position and hierarchy in the network structure. According to the hierarchy, a process running on the leaf node in edge devices, root node in the cloud, and intermediate devices like routers, servers, and so on belong between devices and the cloud. The assignment of work to a process depends on a certain geographical region. All these things are being provided by mobile fog by its API support through its runtime. CPU utilization rate, bandwidth, and so forth are some of the user-given policies which help mobile fog to provide the essential computing instance for the dynamic scaling requirements.

Hong et al. [3] described the different set of event handlers and also highlighted the various set of programming interfaces in the mobile fog framework. Lee et al. [4] described security and privacy issues in terms of fog computing. The main components of fog computing are IoT nodes, fog nodes, and backend cloud, which gives limited resources to compute as well as detection of a malicious attack. "Man in the middle attack" is a famous security attack on IoT networks. In spite of that, detection, as well as prevention, plays the main role in the security of this type of computing. Here, hybrid detection techniques are being used to detect various malicious attacks that are mainly based on signature or behavior taken one at a time or both at a time. Also, all these processes should ensure data integrity in terms of lightweight encryption and decryption algorithms with masking techniques, to become the highest resource user at the time of computation to efficiently secure and protect data.

7.1.3.2 ARM Mbed OS with Pelion IoT Platform

7.1.3.2.1 Arm Mbed Operating System

Arm Mbed OS [5,6] with the Pelion IoT develops an Arm Cortex-M microcontroller for connectivity and security, drivers, and RTOS for sensors and input-output devices. As it consists of a free, open-source embedded OS designed specifically for the IoT and it is now a thread certified component, it gives a minimum-power, self-healing mesh connectivity especially considered for the home, using IPv6 with 6LoWPAN as the foundation. The Mbed OSs supports are devices, real-time software execution, open source, ease of use, an end-to-end security, and drives and support libraries respectively. Figure 7.1 highlights the working coordination of Mbed OS and Pelion device management system [6].

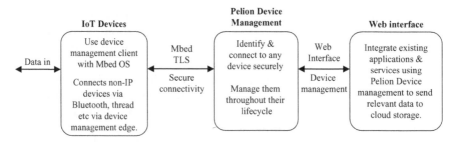

FIGURE 7.1
Operation of Mbed OS and Pelion device.

7.1.3.2.2 *Pelion IoT Platform*

This tool is a secure, flexible, and effective foundation spanning connectivity, device, and information management. It enhances the time to value of IoT deployments by providing easily connected trusted IoT devices on global networks, undetectably administering them, and taking out real-time data to force competitive advantage. Figure 7.2 shows how a device-to-data platform has been built on a strong security framework in Pelion IoT platform.

7.1.3.3 **Open Source for IoT by Eclipse**

Eclipse IoT provides the technology needed to build IoT devices, gateways, and cloud platforms [7].

7.1.3.3.1 *Constrained Devices—Open Source Stack*

Eclipse-libraries are used on a device and eventually give a whole IoT development protocol. The various OSs are as follows: FreeRTOS, Contiki-NG, RIOT, Apache Mynewt, and Zephyr.

- **Hardware abstraction:** Eclipse Edge has board-support packages, libraries, and drivers (provided by silicon vendors) to make a high-level API. The microcontrollers (GPIO, ADC, MEMS, etc.) are well accessed due to this API.
- **Device management:** Eclipse Wakaama has the OMA LWM2M standard.
- **Communication:** Eclipse Paho or Eclipse Wakaama use MQTT or LWM2M like other open-source software.

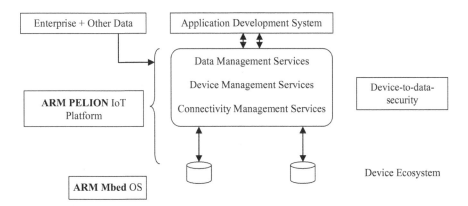

FIGURE 7.2
Device-to-data tool built on a strong security framework. (Courtesy of Home|Mbed, Arm Mbed., Available at: https://www.mbed.com/en/. Accessed August 13, 2018.)

TABLE 7.1

Features and Descriptions of Eclipse Kura Gateways

Features	Description
OS	Linux (Yocto, Core, Ubuntu)
Application-container	Eclipse Conceirge (OSGi runtime), Equinox
Communication & connectivity	API: Serial, ES-485, RLE, GPEO etc.
	Field Protocols: MODBUS, CAN bus etc.
Interfaces—network management	Cellular, Ethernet, WI-fi
Remote data management & messaging	MQTT, Apache Camel message routing engine

7.1.3.3.2 Gateways—Open Source Stack

Eclipse Kura has the ability to observe IoT gateway. Eclipse Kura has a middleware and runtime environment. Eclipse Kura–enabled IoT gateway stack provides the important features shown in Table 7.1 [7].

7.1.3.4 OpenHAB

The openHAB (open Home Automation Bus [8]) is an open-source software that integrates different home automation devices and methods into a distinct solution by using its uniform user interfaces and a common viewpoint to automation rules throughout the whole system, despite the number of subsystems. Generally, it does not need any cloud for data storage but it is cloud-friendly too. The overall distributive architecture of openHAB is shown in Figure 7.3 [8].

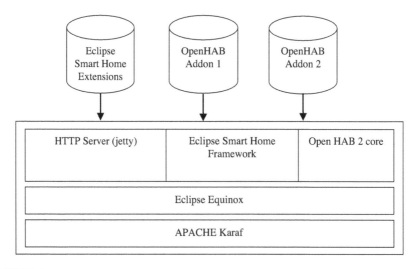

FIGURE 7.3
Overall architecture of openHAB.

7.1.3.5 OpenRemote

OpenRemote [9] is open-source software. It currently has two running versions. The first version, OpenRemote Designer 2.5, is available for private, educational purposes, which enables a cost-effective solution by creating a single installation for the clients to save time. The second version, OpenRemote Manager 3.0, enables the use of service providers: they handle multiple-number user accounts, assets, and parameters like ID service, account management, map navigation, attribute manager, and rules. Figure 7.4 depicts the system model of the OpenRemote framework.

7.1.3.6 Programming Tool—Node-RED

The lightweight programming platform (Node-RED) is applied for assembling APIs, online services, and a hardware device that provides a browser-based editor using a variety of nodes in the libraries by dragging and dropping [10].

Features of Node-RED are the following:

- **Browser-enabled flow editing:** It gives a browser-enabled flow editor. Flows can then be used at the runtime in a single click with drag and drop. JavaScript functions can be created within the editor too. A built-in library helps to save useful functions, templates, and flows for future use.
- **Built on Node.js:** It is non-blocking and an event-driven model. It operates on an edge node such as the Raspberry Pi and also in the cloud. The Node-js package repository is also expandable.
- **Social expansion:** It is easily shareable with an online flow library by importing and exporting throughout the whole world.

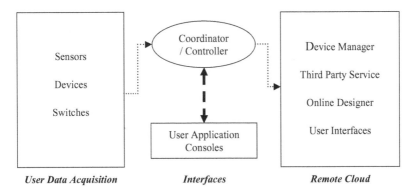

FIGURE 7.4
OpenRemote system architecture.

TABLE 7.2

Different File Types (Erlang)

Type-File	Name-File /Extension	Documentation
Module	*.eri*	Manual reference
File-include	*.hrl*	Manual-reference
File-release resource	*.rel*	Manual rel page in SASL
File-application resource	*.app*	Manual app page in Kernel
Boot script	*.scrint*	Manual script page in SASL
Binary boot script	*.boot*	-
File-configuration	*.config*	Manual config page in Kernel
File-application upgrade	*.appup*	Manual appup page in SASL
File-release upgrade	*relup*	Manual relup page in SASL

7.1.3.7 Erlang Language for IoT (Eliot)

Eliot is a lightweight, open-source framework with the benefit of Erlang-embedded language; it gives an IoT applications simulator for debugging and testing. It is capable of modeling a whole system via virtual nodes by executing unmodified Eliot code in a mixed deployment of nodes whereby virtual nodes continuously interact with the physical devices. Eliot tool provides RESTful interfaces via normal HTTP operation for accessing the users. The following file types are supported by Erlang, which is shown in Table 7.2 [11,12].

7.1.3.8 PyoT

PyoT [1] is a web-based tool designed in the macro programming concept. The programmers can straightforwardly define the test, and share applications via a web interface/IPython notebooks.

This tool distributes processing effort all around the network efficiently. The features of PyoTs are programming multiple numbers of nodes as a whole, sensors monitoring, in-network tasks creation, and network information visualization. PyoT can be tested on real wireless sensor network deployments or in emulated environments effectively. The PyoT architecture [1] is shown in Figure 7.5.

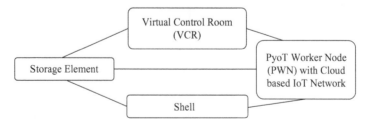

FIGURE 7.5
PyoT architecture.

7.1.3.9 IoTivity

The lightweight, open-source software tool IoTivity [13] is developed by a group of industrialists that is known as the Open Connectivity Foundation (OCF). The two important OCF implementations under the IoTivitys are IoTivity-Lite and active-IoTivity. IoTivity assumes a device with sufficient memory to build a room for every feature of the OCF-1.3 specification. IoTivity is used to implement OCF-control applications for higher-level languages (Node.js/Java). The lightweight IoTivity-Lite is able to focus noded hardware and software platforms and to implement the OCF-1.3 when energy efficiency, resource utilization, and modular customization are required.

7.1.3.10 Dripcast

Dripcast [14] is a new Java framework for cloud-style applications. It provides a very simple way to access Java objects on a cloud. The architecture is based on a distributed processing model, so that cloud objects (that is, data objects on a cloud) will be stored and processed with scale-out style, on unlimited computing resources. It works with a really scalable (scale-out style) cloud platform. The framework changes your development model. The Dripcast tool consists of the following sections:

(a) *Client*: handles the Java object on the devices and forwards calls and that are distracted from the users to the relay;

(b) *Relay*: a stateless distribution-gateway which is familiar with the engine servers and object's unique ID group and which sends requests from clients to consequent engine servers;

(c) *Engine*: every engine server operates JavaVM and executes Java tools of an object for a remote procedure call request transferred by the relay and returns the result back to the relay, and the modified state is stored; and

(d) *Store*: stores Java objects with the potential for automatic recovery and replication.

7.2 Background Frameworks Analysis

From the above discussion, we can analyze some features like the programming approach, language used, and so on for the given programming frameworks. The analysis is shown here in Table 7.3. From Table 7.3 and the discussion of the architecture of the frameworks, it is also evident that the programming approach is divided into several backgrounds itself, as shown in Table 7.3.

TABLE 7.3

Analysis of the Programming Frameworks

Name	Programming Approach	Language Used
Mobile Fog	Macroprogramming	Java, Linda
ARM mbed OS with Pelion IoT	Model-based	Java-script, Java
IoT-Eclipse	Macroprogramming	Java, Java-script
OpenHAB	Database	Java, Python
OpenRemote	Node centric	Java, Python
Node-RED	Node centric	Node-js
Eliot	Macro programming	Erlang
PyoT	Macro programming	Python
Iotivity	Node centric	C, C++, java
Dripcast	Model-driven	Java, rust

7.2.1 Embedded Device Programming

Every electronic device used embedded software [15] today, like DVD players, watches, antilock brakes, mobile phones, toasters, many medical instruments, satellites, and deep-space probes. The embedded software is also used for military purposes for detecting enemy aircraft, guided missiles, and pilot UAVs. According to a 2016 survey by IEEE Spectrum, C and C++ took the top two spots for being the most popular and used programming languages in embedded systems [16]. This is unlikely to come as a surprise to seasoned engineers, scientists, and hobbyists who are almost guaranteed to have used one or both of these languages to a large extent at some point. Also, there is Rust, Python, VHDL, and Verilog.

7.2.2 Message Passing in Devices

The important features of messaging systems are the following: (a) architecture of the messaging model, (b) requirements for data sharing in the message, and (c) architecture of the target system [17]. A summary of the message passing protocols are as follows: AMQP and JMS protocols have been developed for its speed and reliability in transactions for business [18]. JMS is centered on Java systems, but there are a number of proprietary C and C++ JMS-API links which are used in JMS broker. JMS cannot pledge interoperability between producers and consumers even though it uses a different JMS. A simple and lightweight data collection method for all devices is being given by MQTT but it has only partially operated between publishers and subscribers of MQTT. If the format of the message body is not per the rules between peer devices, the message cannot be un-marshaled; hence it is impossible to exchange among different MQTT. REST provides simple client-server interconnections that are necessary for the Internet, but it's not giving good response for asynchronous, loosely coupled, publish-and-subscribe message exchanges. The stateless HTTP model is applied in the server for simplicity; it focuses to make an overhead of extra information [19]. It is very inefficient for request-processing time and consumed resources (e.g., a number of TCP/IP connections). For Internet-based, low-power systems such as wireless sensors, CoAP has been designed. It is mainly used where it is not essential to collect, control, or share data in real time and it is also not necessary for high performances. CoAp is mainly intended to do its work processes in offline mode after the data is captured. It needs an HTTP proxy mapping—that is, CoAP HTTP mapping—to connect it with the cloud back end. The biggest demerit of CoAP is that it is not suitable due to its increasing message latency overhead. Another message-passing protocol is XM-based XMPP. For conference chats or calls, voice and video calls, chatting for more than one person, collaboration among middlewares, syndication of contents as well as general routing, XMPP is being used. Its framework provides service for security (authentication, encryption, etc.) but not for QOS. Its biggest demerit is the overhead of XML parser's additional processing. Therefore, if we go for a comparative discussion about XMPP, JMS, REST, MQTT, or AMQP in terms of systems having processes that need to fan out messages, it requires multi-broker or configuration of servers. But CoAP can multicast IP concurrently by a single request to multiple CoAP devices, which is a great thing itself.

7.2.3 Coordination Model

A coordination model and its languages are used to integrate various heterogeneous components and execute the good features of parallel and distributed systems [20]. The coordinated components are separated based on multilinguality and heterogeneity from the computational components, since the coordination component is a homogeneous way for IPC and abstracts the machine-dependent details. The coordination model provides the use of a heterogeneous entity of architectures. Coordination languages are two types: (a) data-driven—Linda, Bauhaus Bonita, objective Linda, lAURA, Ariadne/hopla, Sonia, GAMMA, Jolie, and ORC; and (b) control driven—Proteus Configuration Language (PCL), conic, Darwin/Regis, durra, CSDL, POLYLITH, RAPIDE, ConCoord, TOOLBUS, and MANIFOLD, respectively.

7.2.4 Polyglot Programming

In computer science, a polyglot [21] is a program or script implemented with the help of multiple programming languages. Here, different languages may perform the same type of command, or the output may be independent, regardless of compilation and interpretation. The two most used methods for a polyglot program are (a) wide use of languages that have different characters for comments and (b) redefinition of different tokens in different languages.

An example of polyglot programming is the following:

```
#define block
#<?php
echo "\010 hello, new polyglot";
$a = 10;
if(($a))
return 0;
?>
#define c
#include<stdio.h>
#define function()
#define true()
void main()
{
printf("hi! Again polyglot!!\n",true);
}
#define c
```

Emacs is an example of polyglot programming [1] that integrates both C and eLisp (a dialect of Lisp). The two major features of polyglot programming are the integration tool and the different programming languages. An inverse pyramid model is used to classify the programming languages in a polyglot model. The first layer of this model is the base layer, which consists of HTML, CSS, Web templating, and SQL and is called the Domain layer. The second layer consists of Groovy, Clojure, Python, Ruby, JS, and so on and is called the Dynamic layer. The third and last layer is called the Stable layer, which consists of JAVA, Scala, and C. As the number of languages used decreases, the architecture of an inverse triangle becomes very vivid.

7.3 Discussion of the Analysis in Terms of Programming Language

In this section, we will discuss the programming languages that are the building blocks of these aforesaid frameworks, methods, protocols, and so on. The main purposes of choosing these languages are to produce a lower runtime overhead as well as a lighter footprint in the time of bit level access to support low-level coding in the inline assembly code. Also, heterogeneous device co-ordination, tolerance of heavy faults at a time due to geographical reasons, and cloud and device interaction co-ordination all play a huge role in choosing these languages one at a time or many at a time, which is in called polyglot.

C is the most used embedded programming language platform [14,22–25]. According to a study, almost 80% of embedded software is being written using C or its different types [26]. The main features of C are lightweight footprint, use of lower-level hardware, availability of more trained C programmers, a short learning curve, adequate support of compilers, and performance. ANSI C provides all these features in its standard. Some of the types of C for embedded programming are nesC, keilC, dymanicC, and B#. The philosophy underlying nesC's are that nesC is an expansion of C, nesC programs are subject to whole program analysis (for safety) and optimization (for performance), a static language, and that it supports and reflects TinyOS's design. The nesC's model provides system design by considering easy-to-accumulate OS-supported applications. The model permits alternate implementations and a flexible hardware/software boundary. This language allows us to write highly concurrent programs on a tool with very restricted resources. Careful restrictions on the programming model, including the lack of dynamic allocation and explicit specification of an application's call-graph, facilitate whole-program analyses and optimizations. Aggressive inclining reduces both memory footprint and CPU usage, and static data-race detection allows the developer to recognize and fix concurrency bugs. When the Keil-C51 compiler is applied, it is called the KeilC language. It has added some new keywords in the scope of C to make it more embedded and system oriented.

Dynamic C is an integrated development system for writing embedded software. It is designed for use with Rabbit controllers and other controllers based on the Rabbit microprocessor. Dynamic C calls the subsequent functions like editing, compiling, linking, and storing in one function and debugging respectively into one program. It has an easy-to-use, built-in, and full-featured text editor. This program can be executed and debugged at the source-generated code level, which makes sure that it supports assembly language to combine both the languages and going separately. The important commands are pull-down menus and keyboard shortcuts that make it easy to use.

B# language is an object-oriented, tiny, and multi-threaded programming tool. It works similar to different languages, namely C, C++, Java, and C# [27]. For supporting modern object-oriented features, it caters to the embedded systems programmer with effective boxing-unboxing conversions, field properties, multi-threading concepts, interrupt handlers, device addressing registers, and deterministic memory defragmenter, which is directly supported by its underlying virtual machine to generate, apply, and reprocess more portable and decoupled software components through different embedded applications. To support these features effectively, B# is linked to its B#EVM. The agent of a stack-based virtual machine permits the embedded systems developer to compose interrupt handlers and to allow device registers in a consistent manner, independent of the underlying target architecture. An example code of B# is the following:

```
class test{
static void main() {
int i = 123;
object o = i; //boxing
int j =(int)o; //unboxing
}}
```

Most embedded programming tools need a few procedures to interface data and device. In contrast, *clixx.io* interfaces directly with an IoT messaging system (MQTT) that provides fast and easy connection to the devices. The less than 50 lines of code are sufficient for IoT programs.

- **JAVA:** The big data and cloud framework can be converged using IoT and M2M that operates from the device to a Java-enabled data zone [28]. Oracle provides a secure, comprehensive, integrated tool for the entire IoT framework across all vertical markets. This language is capable of real-time response for multiple device endpoints, end-to-end security, faster time to market, and integration with IT systems. Java is the better choice and almost 9 million Java developers are present worldwide.

- **Python:** It is the most demanding scripting language [28] tool. The developers are selecting Python when the system provides enough memory and computational power. The Python syntax is clean, simple, and easy to understand for programmers, scientists, and biologists. The MicroPython board and software package is a small microcontroller optimized to operate Python on a small board that's only a few square inches. The Python packages under IoT that are used to write codes are mraa, myseldb, sockets, numpy, opencv, matplotlib, pandas, tinkter, paho-mqtt, and tensorflow, respectively.

- **JavaScript:** According to an Eclipse survey, 41.8% of developers are choosing JavaScript [28], and 31.5% developers are using Node.js. The examples of microcontrollers are Espruino and Tessel, which run on JavaScript. The libraries of JavaScript for IoT include (a) Johny-five, (b) Cylom.js, (c) Jerry-script, (d) Node-MCU, (e) Zetta, and more. A code snippet for Johnny-five is following [29]:

```
var jf = require("johny-five"),b, m,l;
b = new jf. Board();
b.on("ready",function(){m = new jf. Motor({pin:5});
b.repl.inject({m`: m});
m.on("start",function(){
console.log("start",Date.now());
b.wait(4000, function(){,.stop();})});
m.on("stop",function(){console.log("stop",Date.now())};);
n.start();});
```

- **Linda [30]:** It is used to coordinate and communicate among various objects that can be stored and retrieved from the memory. Linda is an implementation of "coordination language" in which "tuples" on the data objects are then associated with a C and memory (tuple space). This model needs major operations like (a) consuming a tuple from tuple space, (b) reading a tuple space non-destructively, (c) writing into a tuple space, and (d) creating new processes to evaluate tuples, writing the result into tuple space.

  Examples of LINDA Programming [30]

Code	Explanations
out("sum", 2,3)	Add tuple into tuple space and immediate process.
in("sum",?i,?j)	Assigns 2-to-i, 3-to-j and tuple separated from the tuple space.
rd("sum",?i,?j)	Assigns 2-to-i, 3-to-j and tuple not separated from the tuple space.
eval("ab",-6,abs(-6))	Create process to compute the absolute value of -6. Active tuple resolves into the passive tuple ("ab", -6, 6), then is read by an in or rd.
eval("roots",sqrt(4), sqrt(16))	Creates a live tuple. Numeric outcomes combine to form a three-element tuple saved in tuple-space.

- **Java Orchestration Language Interpreter Engine (Jolie):** Jolie is a micro-services-based, open-source programming language [31]. In the programming paradigm of Jolie, each program can interact with different programs by processing messages. It gives an abstraction layer that permits services to interact via various mediums, from TCP/IP sockets to local in-memory processes. Some examples of compositions are orchestration, which offers operations obtained by coordinating other services; aggregation, which is used to create APIs of separate services; redirection, which hides the real service locations to clients by handover logical names; and embedding, which operates other services as inner apparatus. Jolie gives fast communications and can even operate code written in different languages (Java and Javascript).

- **ORC:** This language is developed to provide concurrent-distributed tools and is also simple and sensitive to write [32]. ORC is well known for process-orchestration, a structure of concurrent platform work-flow applications, web-service orchestration, and business process management, which gives a flow of control and concurrent systems. The complex ORC expressions are used for simpler expressions that are not publishable.

  A simple code snippet in ORC is as follows:

```
{- Publish 1 and 2 in parallel -}
(0 | 1) >n> n+1
{-
OUTPUT: PERMUTABLE
12
-}
```

7.4 Future Scope

In the near future, it is believed that if IoT-cloud is a unified model, it will be a powerful platform for the programmers' as well as the commoners' approaches. According to Cisco [1], the current internet-enabled intelligent devices may be 2% of the total devices, but more and more devices are connecting every day, in large number, to the Internet itself. Therefore, the simple and supportive programming frameworks are more demanding. The following problems also need to be considered: (a) similar to CMM, a standardization of the frameworks should be maintained; (b) a lighter version of the high level programming languages must be evolved efficiently; (c) the need to standardize also in the application layer protocol of IoT must be addressed; (d) standardization of legacy programs; (e) enhancement of polyglot programming; (f) introducing artificial intelligence into devices to authorize reliable and secure frameworks for application development; and (g) simplification IoT application using lightweight GUI usage.

7.5 Conclusion

A survey and review of some of the important features for the IoT programming paradigm have been discussed in this chapter briefly. This chapter expresses the programming techniques and languages in some of the models that have been the latest in the trade. There are lots of programming approaches in the programming paradigm, like imperative, object-oriented, declarative, and so on. By going through the above chapter, one can easily understand that the whole paradigm of IoT programming is a mixture of the good things from every known paradigm technique. C (mainly embedded versions of C) is now too the most used language in the IoT programming paradigm, which is imperative in nature and does not have the reusability feature. Also, object-oriented languages like JAVA, Python, JOLIE, ORC, and so on have their fair share in the programming paradigm of IoT with their microprogramming, orchestration, and coordination abilities. This is why we can easily conclude that IoT developers need to make their own standard programming platform by using all these approaches, which itself is challenging work for IoT programmers. As for a new domain, the standardization of models in the future will need regular attempts from the IoT developers too. Therefore, the essential features of programming paradigm of IoT can be concluded as (a) message-passing frameworks, (b) features of embedded language, and (c) the importance of different languages in polyglot programming, all of which can be deduced from the above chapter. Also, the main aim of this chapter is not only the comparison of the programming languages with respect to different IoT programming platform, but also to give a comparative picture and narration for the future programmer community to make an effective and concrete standardization of programming structure in the IoT programming paradigm.

Abbreviations List

6LoWPAN—IPv6 over Low-Power Wireless Personal Area Networks

ADC—Analog to Digital Converter

AMQP—Advanced Message Queuing Protocol

API—Application Program Interface

CoAP—Constrained Application protocol

CSS—Cascaded Styled Sheet

DHT—Distributed Hashtable

GPIO—General Purpose Input Output

HTML—Hyper Text Markup Language

IM—Instant messaging

IoT—Internet of Things

Js—Javascript

LWM2M—Lightweight Machine to Machine Environment

MEMS—Micro-electromechanical systems

OMA—Open Mobile Alliance

OS—Operating System

OSGI—Open Server Gateway Initiative

REST—Representational State transfer

RTOS—Real-Time Operating System

XML—Extensible Markup Language

XMPP—Extensible Messaging and Presence Protocol

WSN—Wireless Sensor Network

References

1. Buyya R., Dastjerdi A. V., *Internet of Things, Principle and Paradigms*, Elsevier, 1–380, 2016.
2. Rahman L. F., Ozcelebi T., Lukkien J. J. Choosing your IoT programming framework: Architectural aspects. *IEEE 4th International Conference on Future Internet of Things and Cloud*, Los Alamitos, CA: Conference Publishing Services. 2016. doi:10.1109/FiCloud.2016.49293.
3. Hong K., Lillethun D., Ramachandran U., Ottenwälder B., Koldehofe B. *Mobile Fog: A Programming Model for Large-Scale Applications on the Internet of Things.* Hong Kong, China: ACM, 2013.
4. Lee, K., Kim, D., Ha, D., Rajput, U., & Oh, H. (2015, September). On security and privacy issues of fog computing supported Internet of Things environment. In *Network of the Future (NOF), 2015 6th International Conference* on the (pp. 1–3). IEEE.
5. The Three Software Stacks required for IoT Architectures. Available at: http://Ioteclipse.org, latest update, 2017.
6. Home | Mbed., Arm Mbed. Available at: https://www.mbed.com/en/. Accessed August 13, 2018.
7. Eclipse Leshan: Eclipse RSS. Available at: http://www.eclipse.org/leshan/. Accessed August 13, 2018.
8. Empowering the Smart Home. OpenHAB. Available at: https://www.openhab.org/. Accessed August 13, 2018.
9. Community: OpenRemote. Available at: http://www.openremote.com/community/. Accessed August 13, 2018.
10. Node-RED. Available at: https://nodered.org/, Accessed August 13, 2018.
11. Erlang Programming Language, Overall RSS 20. Available at: https://www.erlang.org/docs. Accessed August 13, 2018.
12. About the Nest API | Nest Developers, Nest Developers, Available at https://developers.nest.com/documentation/cloud/get-started, Accessed August 13, 2018.
13. Home: IoTivity. December 18, 2017. Available at: https://iotivity.org/. Accessed August 13, 2018.
14. Dripcast: Dripcast Top. Available at: http://dripcast.org/, Accessed August 13, 2018.

15. Diakopoulos N., Cass S., Interactive: The top programming languages 2016. Available at: https://spectrum.ieee.org/static/interactive-the-top-programming-languages-2016.
16. B#—A Programming Language for Small Footprint Embedded Systems Applications: Part 1, Embedded. Available at: https://www.embedded.com/design/prototyping-and-development/4006620/B, Accessed August 13, 2018.
17. Adlink: Messaging Technologies for the Industrial Internet and the Internet of Things Whitepaper, A comparison between DDS, AMQP, MQTT, JMS, REST, CoAP and XMPP. Available at: https://www.adlinktech.com, 2017.
18. Montesi F., Guidi C., Lucchi, R., Zavattaro G., JOLIE: A Java orchestration language interpreter engine. *ENTCS*. 181, 1–15, 2006. Available at: http://www.cs.unibo.it/cguidi/Publications/Coorg06.pdf.
19. Author: Language Project, Orc Language, https://orc.csres.utexas.edu/, Accessed August 13, 2018.
20. Papadopoulos G. A., Arbab F., *Coordination Models, and Languages.cwi*, Centre for Mathematics and Computer Science, Amsterdam, The Netherlands. Available at: https://doi.org/10.1016/S0065-2458(08)60208-9, 1998.
21. Polyglot (computing), Wikipedia. August 5, 2018. Available at: https://en.wikipedia.org/wiki/Polyglot, last modified August 13, 2018.
22. Dynamic C 32 v. 6.x Technical Reference, Part Number 019-0083. 020330—B, USA, 2002.
23. Dynamic C User's Manua0, Digi International® Inc., 2011.
24. Keil Embedded C Tutorial. DC Motor Interfacing with Microcontroller Tutorial. Available at: https://www.8051projects.net/wiki/Keil_Embedded_C_Tutorial, Accessed August 13, 2018.
25. Gay D., Philip L., Robert V. B., NesC: A Programming Language for Deeply Networked Systems. Available at: http://nescc.sourceforge.net/, Accessed August 13, 2018.
26. Buyya R. and Dastjerdi A. V. (2016). *Internet of Things, Principle and Paradigms*, edited by Krishnamurthy J. and Maheshwaran M., pp. 79–102.
27. Apple Inc. "HomeKit." Purchase and Activation—Support—Apple Developer. Available at: https://developer.apple.com/homekit/, Accessed August 13, 2018.
28. Zola, A., 12 Popular Programming Languages for IOT Development, Hiring | Upwork. May 21, 2018. Available at: https://www.upwork.com/hiring/for-clients/programming-languages-for-iot-development/, Accessed August 13, 2018.
29. Shovic, J. C. Sensing your IOT environment. *Raspberry Pi IoT Projects*, 9–61, 2016. Accessed August 16, 2018. doi:10.1007/978-1-4842-1377-3_2.
30. Programming Examples: Linda—Programming Examples. Available at: http://programmingexamples.wikidot.com/linda, Accessed August 13, 2018.
31. Dmitrii: "Erlang Is Dead. Long Live E...?—Dmitrii—Medium." Medium. February 19, 2016. Available at: https://medium.com/@dmitriid/erlang-is-dead-long-live-e-885ccbcbc01f, Accessed August 13, 2018.
32. IoT Framework, Clixx. IO: IoT Framework. Available at http://clixx.io/iot.html, Accessed August 13, 2018.

8

Basics of the Internet of Things (IoT) and Its Future

Neeraj Sharma, Vijender Kumar Solanki, and J. Paulo Davim

CONTENTS

8.1 Introduction

This part of this chapter gives an overview of IoT and also explains the history and growth of the components of IoT and its applications.

- **IoT Applications**: There are numerous applications of IoT, ranging from home automations to agriculture, defense, and many more.
- **Components Related to IoT architecture**: There are a number of components in IoT like sensor devices (thermal, humidity, pressure, etc.), network technologies (Wi-Fi, Wimax, cellular, Bluetooth, etc.), protocols and standards, isntelligent analysis, intelligent actions.
- **Future of IoT**: As the technology of IoT is improving day by day, the increased easy access to the Internet makes the future of IoT very bright now. Google search trends show that the comparative growth of IoT is tremendous as compared to other technologies.

The Internet is defined as an interconnection of various networks that are operated by governments, academia, and industry. In the past three decades the Internet has continued on an expansion graph in terms of users and infrastructure. According to a survey by Internet World Stats [1], more than 4.156 million people use the Internet for browsing various types of applications, with content ranging from media files and images to music files and video files. Online gaming is also very popular among Internet users. A growing number of Internet users is due to many reasons, ranging from housewives looking for cooking videos to businessmen handling complex business processes through online modes. Moreover, getting access to business data while sitting anywhere across the globe and at any time is one of the most beneficial virtues of the Internet. One can be virtually present in an office to observe all the routine activities by using the Internet.

Furthermore, the Internet of Things (IoT) is the latest revolution, which acts as a global platform for connecting cyber physical devices and machines to communicate, exchange data, and coordinate among each other for different tasks. According to a survey, 9 billion devices are connected [2] to the Internet.

Kevin Ashton used the term "IoT" for the first time in 1999 while working on a project of supply-chain management. Moreover, the first written and refer able source available that mentions the IoT was found to be the white paper published in November 2001 at MIT Auto-ID Center [3].

In general, the IoT is the network connection of heterogeneous cyber physical devices, electronic gadgets, sensors, home appliances, software, and actuators that enable data sharing and processing. Each device is uniquely identified by a computing system and is able to interoperate using Internet infrastructure.

The phenomenon of IoT contains two parts: "Internet" and "Things," whereas the latter, "Things," refers to those intelligent devices, sensors, and actuators, devices that are able to communicate with other devices, whenever required, thereby making things accessible anytime without restriction.

The IoT is built on three pillars [4] related to ability of smart objects:

1. They should be able to be identifiable uniquely (anytime and anywhere).
2. They should be able to communicate (anywhere at any time).
3. They should be able to interact with each other, or on the Internet (anytime and anywhere, with any node).

The Internet of Things permits objects' processes to be controlled remotely using available communication infrastructure. This involves integration of physical devices with computer-based systems, resulting in efficient, accurate, and economical system mechanisms with much less human intervention [5].

8.2 Applications of IoT

The applications range in various domains like home, defense, agriculture, manufacturing, and many more. The IoT is playing a major role in saving the lives of patients in smart healthcare, where the access to medical professional

IoT	Home	Fans, fire alarm, HVAC, health monitoring and exercise guidance
	Defense	Surveillance, equipment tracking, threat analysis, troop movement, intrusion detection, landmines handling, drones
	Farming	Soil analysis, livestock monitoring, fleet, seed analysis, water and fuel storage, waste management
	Traffic Management	Traffic flow monitoring, congestion control, smart parking
	Remote Monitoring	Environ monitoring, health monitoring, temperature monitoring in furnaces
	Public Transport and Logistics	Toll tags, lights, real time routing and analysis, usage based billing, RFID and NFC supply chain management, inventory management.
	Entertainment	Virtual gaming consoles, digital camera digital TV
	Healthcare	Medical implants, wearable vital trackers, PDAs, telemedicine
	Utility and Public Services	Security monitoring, transportation, energy billing, POS terminals
	Security	Homeland security, border security, camera surveillance
	Office Industry Automation	Production flow, inventory monitoring, attendance, entry restriction.

FIGURE 8.1
IoT application areas.

is possible through telemedicine. The IoT is a having vast area of applications ranging from home to everywhere you can imagine; some of the application areas of IoT are as given in Figure 8.1.

8.2.1 Components Related to IoT Architecture

Another IoT basic architecture is comprised of various sensors, IoT gateways, Internet, cloud server, and mobile IoT applications. In Figure 8.2 we have seen that different types of sensors are there; these sensors collect the sensor data

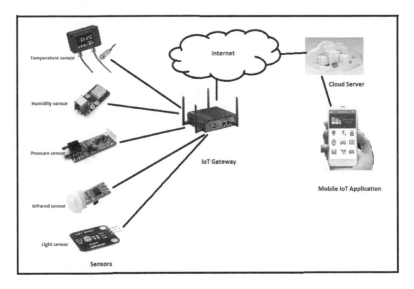

FIGURE 8.2
Basic components of IoT.

and send the data to the cloud through an IoT gateway, which in turn is connected to the cloud through some kind of communication medium. The cloud contains the relevant information, which is to be used by the system to perform certain tasks as per user instructions through IoT mobile applications.

Broadly, the Internet of Things elements can be divided in to five categories [6], which have been defined by the big IT company Cognizant Technology Solutions working in the Internet of Things:

1. Sensor devices (thermal, humidity, pressure, etc.)
2. Network technologies (Wi-Fi, Wimax, cellular, Bluetooth, etc.)
3. Protocols and standards
4. Intelligent analysis
5. Intelligent actions

8.2.1.1 Sensors

In IoT sensors [7] are small electronic hardware devices that produce electrical signals or digital data, which is derived from a physical state or event. This data is converted by other devices into information used for smart decision making by its user and or intelligent devices. There are mainly fifteen types of sensor devices used in the Internet of Things platform for different applications:

1. **Temperature Sensor [8]:** This is an electronic device used to determine the amount of heat energy or even coldness that is generated by a solid/liquid or system, allowing us to detect any physical change in temperature by its producing either an analog or digital output.
2. **Humidity Sensor [9]:** The humidity sensor, also called the hygrometer sensor, is used to measure the relative humidity in the air. It can measure both moisture and air temperature of the system.
3. **Pressure Sensor [10]:** A pressure sensor is an electronic device used to measure the pressure of the gases or liquids and convert it into an electric signal. A pressure sensor is basically a transducer where the signal strength is proportionate to pressure applied.
4. **IR Sensors [11]:** An infrared (IR) sensor is an electronic instrument that is used to detect a selected light wavelength (between 0.75 and 1000 μm) in the IR spectrum of its surroundings by either emitting or detecting infrared light.
5. **Image Sensors [12]:** Image sensors are electronic/Complementary metal–oxide–semiconductor (CMOS) sensors used to convert optical images into digital signals to display or store files in digital form.
6. **Optical Sensor [13]:** An optical sensor device is used to convert light rays into electrical signals, which can be easily readable by an electronic instrument/device.

a. **Photodetector**: This uses a light-sensitive, semiconductor material–based p-n junction photodiode, which in turn converts light photons into electrical currents, as the case in devices like photodiodes or phototransistors.

b. **Pyrometer**: This device is used to estimate the temperature of an object by sensing the color of the light emitting from the object. The color and temperature are compared to reference temperature and color that the object emits.

7. **Gyroscope Sensors [14]**: A gyroscope sensor can be used to provide information related to orientation and rotational movement, also called angular movement. It can be used to measure rotational speed of an object around its axis.

8. **Accelerometer Sensors [15]**: An accelerometer is an electronics device (transducer) that senses the earth's gravitational acceleration in the device; the sensor may be installed in it and convert it into electrical signal, which in turn is converted to digital information for use in decision making. These devices are used in image orientation in mobile phones.

9. **Motion Sensor [16]**: Motion sensors are devices that are used to detect the movement of any type of object (living/nonliving). These devices usually detect the reflection of light by that object. These are passive infrared, microwave, ultrasonic, and others.

10. **Level Sensor [17]**: A level sensor, as the term itself suggests, is a sensor used to detect the level of liquid, oil, or liquefied solids in an open system or in a closed container. Level sensors are used in automobiles for fuel level indicators and empty fuel tank warning systems.

11. **Smoke Sensor [18]**: A smoke sensor device is used to detect the presence of smoke in a room or environment system. The detection is done by either photoelectrically or by ionization.

12. **Gas Sensor [19]**: Gas sensors are used in industries to detect the leakage of harmful gases in a hazardous system. These gas sensors can detect the presence of flammable or toxic gases.

13. **Chemical Sensors [20]**: Chemical sensors are a combination of many types of sensors—for example, ion sensor, gas sensor, bio sensor, semiconductor gas sensor, or enzyme sensor—which in turn are applied to analyze the chemical composition and concentration of a given chemical.

14. **Water Quality Sensors [21]**: Water quality sensors are used in water distribution systems that monitor the ionization of water and quality of potable water. Different types of water quality sensors are the list that follows:

a. **Chlorine Residual Sensor**: This measure chlorine residual in drinking water, mostly used disinfectant for potable water.

b. **Total Organic Carbon Sensor**: The water may contain organism with organic properties. The presence of organic carbon elements in the drinking water is measured by these sensors.

c. **Turbidity Sensor**: In rivers and streams there are a lot of solid particles suspended in the water. The presence of these solid suspended particles is measured by turbidity sensors: applications include water stream gauging and effluent measurement in wastewater.

d. **Conductivity Sensor**: The presence of dissolvable solvents changes physical properties, including conductivity of liquid. The conductivity sensors are used to measure the presence of dissolved compound solvents in industrial wastewater.

e. **pH Sensor**: The pH sensors are used to know whether the given liquid chemical (or water) under testing is a base or an acid.

f. **Oxygen-Reduction Potential Sensor**: This sensor is used to gauge whether the given solution has oxidizing or reducing features.

15. **Proximity Sensor [22]**: The proximity sensor gives an idea about the presence of some object with precise distance information without getting in touch with the object.

8.2.1.2 Network Technologies

Signals assimilated by sensors are transmitted over communication networks with all the network components, like network switches, routers, and bridges in different network topologies: say, LAN, MAN, and Wide Area Network (WAN). The network should be able to uniquely identify each sensor node attached to the system. Different communication technology may affect the hardware requirements and cost of the system. There is a range of network technologies for a system of the Internet of Things to communicate with the nodes connected to it.

- **Wi-Fi [23–25]**: Wi-Fi is a reliable wireless infrastructure for communication of digital data in the current time. This technology is given priority for IoT. This was developed to give freedom from Ethernet and to go wireless. It is a short-range wireless connectivity solution, with interpretability features. Some of the features of WiFi include low cost of devices, ease of deployment, point-of-presence, high power consumption, congestion, and moderate range for communication.

- **Cellular [26]**: This application is IoT-based and requires operations over long distances using cellular (GSM/3G/4G) communication networks for data sharing. Now the cheaper rates of data packages have increased the capability to send huge amounts of data, especially for 4G; it is ideal for sensor-based, low-bandwidth-data projects that will send very low amounts of data over the Internet. Cellular technologies are designed for reliability, security, and scalability and can provide a strong foundation for IoT connectivity with a unique combination of functionality and good performance. Small software upgrades and existing deployed infrastructure provide a base for unmatched IoT coverage and fast time-to-market. Cellular IoT is leading the 4G transformation; 4G is build-on and extends the cellular networks, bringing new levels of performance and functionality to the networks and enabling new IoT services, ecosystems, and business models.

- **Bluetooth [27]**: This is a short-range communication technology. In 2014 Bluetooth 4.2 was announced with a combined basic data rate and low power requirement for RF transceivers. In 2016 Bluetooth 5 was announced with many added features, which include increased range, faster data transfer speed, and bigger message sizes that it can transmit to other devices. Bluetooth 5 has improved the IoT implementation experience with easy implementation and effortless interactions among all connected devices.

- **Zigbee [28]**: The unique selling point of Zigbee is its low-cost solution for wireless area networks (WANs), consuming much less power, which is the biggest worry for engineers who are designing the IoT with battery-based devices for monitoring and controlling the system. The Zigbee chips can be integrated with microcontrollers in IoT applications. There are three Zigbee specifications, as follows:

  - *Zigbee-Pro* [28]: The main specifications of Zigbee-Pro are its capability of cross-band communication across 2.4 GHz frequency band and its ability for device discovery and pairing. It is able to unicast, multicast, and broadcast messages with security in place (AES-128).

  - *Zigbee-IP* [29]: The Zigbee-IP is open standard, working on IPV6-based wireless communication and providing mesh topology network solutions. It uses internet for seamless connections to control low cost and low power devices. It can connect many devices with flawless interaction.

  - *Zigbee-RF4CE* [30]: The Zigbee-RF4CE is designed to give solutions for two-way device-to-device communication with remote control commands. It is a low-cost, small-size device.

- **Z-Wave [31]**: Zensys, a Danish company based in Copenhagen, has developed the Z-wave. Z-wave is a wireless transmission protocol mainly used in IoT applications. Z-wave uses low-energy radio waves for communication among devices over a wireless mesh network. It gives freedom of interoperability of different hardware devices and software. Z-wave devices can communicate with each other directly and indirectly through adjacent device nodes. They can communicate within a range of 30 meters at a permissible bandwidth range of 908.42 MHz in North America and 868.42 MHz in Europe.

- **6LoWPAN [32]**: As the full name implies—"IPv6 over Low-Power Wireless Personal Area Networks"—LoWPAN is a wireless mesh network topology used for connecting IoT devices to the cloud. LoWPAN truly gives a solution of a low-power IP-based, nod-connected network, which uses IPv6 data packets to be transmitted over the network, making communication fast, efficient, and robust. Its IP-based structure takes advantage of open standard research for application in the IoT.

- **Thread [33]**: The idea of Thread communication technology was conceived by a Thread group in 2014. The Thread is designed as a replacement for WiFi in IoT applications, as WiFi has many limitations when we use it in IoT applications, like limited node connecting capability as compared to Thread, in which more than 200 nodes can be connected. Thread is IEEE802.15.4 IPv6 standard. Communication is IP-based, supporting wireless mesh topology.

- **Near-field communication [34,35]**: The NFC is a high-frequency, short-range (10 cm) mode of wireless communication for data exchange from close-proximity devices. Being short-range communication, it is safe and secure. Device pairing is very fast, need much less power in applications.

- **SigFox [36]**: SigFox gives tailor-made solutions for long-range communication for the IoT, enabling remotely located devices to connect using ultra narrow band (UNB) technology. It needs much less energy for transmission of data. This network has the capability to send up to 140 messages per day on a data speed of 100 bits per second. This technology uses ISM radio band at 915 MHz in USA. The distance for communication in rural area is about 50 Km and 10 km in urban area, less as compare to rural due to noise and obstructions.

- **Neul [37]**: Neul operates in sub-1 GHz band. It uses small slices of TV white space available for use, giving chance for higher scalability and coverage as well. This is cheaper, low-power solution for wireless communication on a small chip, called an Iceni chip, uses UHF band, at present not being used. Modern digital TV technology does not use the UHF band. Neul technology uses much less power and the device needs small batteries to operate; transmission range is 10 Km at 100 kbps.

- **LoRaWAN [38]**: The LoRaWAN is low-power-consuming, working on small batteries, for powering a wireless area network. It can be used anywhere as communication for local nodes and regional, national, or international networks with IoT-required key features like end-to-end secure bidirectional communication with mobility. This technology provides seamless interoperability among different device nodes.

- **Long Term Evolution (LTE):** LTE [38] is a standard for high-speed wireless communication used in the IoT when larger amounts of data are to be transmitted among devices. This network standard was designed to use an IP-based system of communication for reduces data transfer latency. LTE was first proposed in 2004 by DoCoMo, a Japanese telecom giant. It supports fast mobility and multicasting and broadcasting.

- **WiMAX [40]:** Worldwide interoperability and Microwave Access is a name coined by the WiMAX Forum in 2001. This technology was designed as an alternative to cable broadband and DSL for last mile wireless broadband service. WiMAX can transfer data higher than 1Gbits/sec. It can provide communication services like IPTV and VoIP.

8.2.1.3 *Protocols and Standards [41]*

In IoT we generate so much data through sensors that this available data needs to be aggregated and analyzed for use by existing technology and for deciding the action to be taken by the attached devices. Only standardization can ensure the best desired results. If these standards are not followed, the outcomes can be adverse and different than the expected results.

An RFID-based tag system standardization has been done in last years. This standardization process is limited to sensor/RFID only. For RFID, ISO 28560 specifies encoding standards and data modeling [42]. Near Field Communication Interface Protocol (NFCIP) [43] has been already standardized (ISO 18092, 21481 and ECMA 340/352). The biggest issue in the Internet of Things is in its naming system: as most IoT applications expect unique identities for each device, a global coordination for standardization is the need of the hour. So in terms of communication technology, though we have developed IEEE standards for almost all applications, in machine-to-machine (M2M) communication we need to be precise while proposing standards for M2M communication with unique identities for devices.

8.2.1.4 *Intelligent Analysis*

An intelligent analysis application takes an input data feed from various IoT sensor devices in real time; it converts the data into clear information for human analysis and use. The system analyzes the information received from sensors as output, resulting from some specific condition and setting. It helps the industry in automation by providing vital industry result analysis.

In the IoT, data is analyzed in real time; this real time does not mean fast processing but is about guaranteed response. True real-time application analysis means response within a stipulated time with limited, well-defined latency.

Data processing can be done in two way:, one is at the node (sensor) side, called edge processing, but it has limited processing capability and power-supply related concerns as well. The second method is by sending the sensor data to a processing facility through communication means and storing the data on the cloud, then processing it at the server. It has transmission delays associated with it. There is a new approach, now called the Edge cloud, in which the sensor data is stored on smart devices like smartphones having storage capability.

This is the most important component of the Internet of Things when you have deployed a very good sensor: if we are not analyzing the data smartly and in real time, then we are not taking action if required and the whole system will be a big failure. So we need analytical software as well, which can trigger the event on analysis of some event in an IoT application to work correctly. For example, if you have installed a motion sensor for a camera to capture images in the jungle, the camera should start capturing images only after the motion detection, so an event of motion detection should trigger the image capture in your digital camera. So in IoT applications, analysis of sensor data should be in real time so as to give expected results.

8.2.1.5 Intelligent Actions

Assume a situation if a fire (smoke) detector has detected the smoke in the system but the connected fire alarm is not activated, or a temperature sensor in a car is sensing the car temperature but your auto climate control AC is not activated in winter. These are two examples that can help you in understanding the importance of intelligent action triggered by sensor data analysis. In this phase of IoT implementation, we put action trigger/switching devices that must start/stop some activity in the system as a result of information received from sensors. These actions should be smart actions. This is most important component, which should give the output expected by the user. If this fails then the whole system may fail.

8.3 Challenges in IoT

The Internet of Things is beyond the earth-connected devices that provide key physical data for further processing that data on the Edge or on the cloud to deliver business insights. Many companies are focusing themselves on the IoT and connectivity of their future products and services. For the IoT industry to thrive in a big way, it has to overcome the challenges that have emerged due to many concerns like security, privacy, connectivity, interoperability, software, trust, standards, and many more yet to come into the picture.

- **Security**: Security is the most significant challenge in the IoT. It has been considered as an essential pillar for the IoT. Exponential growth in connected cyber-physical devices has attracted attackers to exploit security vulnerabilities [44,45] in the poor device design, which can lead to user data theft and exposure. Inadequate protection of devices against data security can be catastrophic for the user. There are three key considerations in security: data confidentiality, privacy, and trust.
  - **Data Confidentiality**: This guarantees that only authorized user can access, use, and modify data. In the IoT, data is used by not only by users but also by some devices. So we have to guarantee that data is to be used in a controlled manner and that all the object entities are to be managed in such a manner that they can be authenticated properly. So we need to keep some access-control mechanism in place. So in nutshell, we have to ensure that data should be authenticated, confidential, and integrated while being transmitted or stored. All devices should also be uniquely identified, and one identity management system should be in place to manage the identities of all connected devices.
  - **Privacy**: Privacy is a main concern in IoT applications for healthcare, in which the user has apprehensions about disclosure of identity and data. In the IoT there is no concrete mechanism to ensure the privacy of the user, and user-related personal information created the need for technology for privacy. Most of the data being transmitted through wireless communication poses a privacy violation threat. In such conditions the attacker has access to a network wirelessly. Some frameworks have been proposed like Non Functional Requirement (NFR) [46] Goal-Based Requirements Analysis Method (GBRAM) [47], Kaos [48], Privacy Safeguard PriS [49], and others. The UML conceptual model presents a functional module for enforcing privacy policies through high-level abstraction suitable for IoT applications requiring different levels of privacy. There should be rules/policies related to data governance and user privacy concerns; when users' identities are in the system, they should get assurance of privacy related to sensitive data.
  - **Trust**: The word "trust" [50,51] itself has a diverse meanings with complex ideas; it is highly important, but a consensus is absent in computer literature regarding trust. Trust cannot be established through any matrices or evaluation methodologies. Trust basically

refers to a security policy regulation, controlling the access of resources/services where access of resources/services can be gained through user credentials only. In the IoT, trust analysis is conducted before sharing credentials. Trust negotiation is based on mutual interactions and repeated credential sharing, successfully sharing digital credentials only after verification of properties, which can make the user/device eligible to access the resources/services. So a trust negotiation system uses digital identity information for granting controlled access to resources and services.

- **Connectivity [52,53]:** In the IoT system, connecting a large number of devices will be the biggest challenge. Different applications have different communication models and technologies. We are using a centralized model for authentication, authorization, and connection establishment among different nodes of the network. This model may work efficiently while the number of nodes is smaller, but if the number of nodes is in the tens of thousands, then we have to opt to use the cloud for big data exchange among devices, but maintaining a cloud server is a costly option with a probability of the server going down. Decentralization is a good option for the IoT and implementation of Edge. Using fog computing involving smart devices to handle the critical operations and data clouds for data collection and analytics. Peer communication is simple to implement, where participating devices identify and authenticate each other and start data exchange without involving any administrator, but security is a challenge in such data exchange.

- **Interoperability:** In the IoT, interoperability [54] is matter of concern because consumers may use devices coming with different network interfaces and technologies, developed by different vendors. These devices should be easy to connect and have easy-to-use features to work together. They should interoperate to jointly execute the processes. So interoperability is a capability in system devices, which makes system services and applications work together to give expected results with full reliability. WiFi, Zigbee, Bluetooth, and Z-wave technologies provide a wide range of interoperability, with some groups behind them to provide reliable solutions like the WiFi Alliance.

- **Self-Management:** Self-configuration [55] takes place when some change in configuration is detected. In the IoT many devices are connected to share information with the user that is related to the well-being of devices, so that the user can decide upon corrective measures to avoid shutdown in critical systems. In sensor nodes there are issues related to environment; then these nodes need to self-manage, without human intervention, such as in automatic addressing in Zigbee.

- **Power Consumption**: The devices connected in the IoT applications must be efficient in power consumption and optimized for operating to get maximum battery life while communicating; distance should be kept in mind [56]. In the case of CPU usage, data transfer should be optimized to reduce the battery power drainage.

- **Maintainability**: IoT devices must be designed with a focus on maintenance of devices with a system of ranking of erroneous behavior [57] and repairs of various devices easily, quickly, and cost-effectively. In the IoT environment node failure, connection loss and battery exhaustion must be countered with some alternative measures in the system or with backup plans.

- **Bandwidth**: Bandwidth is a big constraint in IoT applications as numerous devices are communicating with each other and sharing the total bandwidth, which should be less than the Gateway bandwidth [58]. Bandwidth needs for each device are different due to different data generation's buffer and bandwidth allocation for different devices. This problem becomes cumbersome in the case of video streaming from different devices (video streaming needs a wide frequency band).

- **Identification**: Identification of all devices in the IoT is also a big challenge, as all devices should be identified in a secure manner in case of security-related applications. Identification can involve Quick Response (QR) coding or with the object description itself. Scalability of device addressing must be sustainable [59]. The addition of any number of new devices and nodes must not be a constraint.

- **Resource Management**: In IoT applications, efficient resource management [60] is a critical issue, as computation is heavily affected in case of network congestion and latency, causing more energy drainage. Edge computing is a solution for these conditions. We have to look for cheaper and feasible options. The options of resource management include auction-based resource management and optimization of resources. In auction-based resource management, the user may bid for resource use and the service provider may get incentive for maximizing the use of the resource. This concept is used by data centers and cloud service providers. In case of optimization, an application may handle resource allocation and cost (to get profit).

- **Storage**: In the IoT a lot of sensors and other devices generate bountiful data [61], and it is tough to process and store the data in-house. We need third-party servers outsourced to storage service suppliers. Involvement of a third party increases the risk of data integrity and reliability. A resource access control mechanism can be applied for protection of data.

There are some other issues, like unauthorized access to RFID, leading to the possibility of confidential information exposure. RFID tag impersonation, RFID password decoding [62,63], RFID virus [64], and SpeedPass hacking need to be taken care of. As the sensor nodes are scattered on different locations on the network, node security is a challenge: many types of attacks can be initiated on them, like jamming the frequency (which results in no data transmission), and Sybil attacks with multiple pseudo-identities for a node, creating confusion of identity. Flooding attacks can result in data traffic choking and memory exhausting.

In the cloud environment, man-in-the-middle attacks, malicious insider attacks, and hijacking are very common in cases of unencrypted data. So, proper care is needed.

8.4 Future of IoT

As the technology of the IoT is improving day by day with increased easy access to the Internet, the future of IoT very bright now. The youth of Gen Y are more adapted to use IoT-integrated devices capable of monitoring their daily activities and are looking for more IoT application–based devices. The IoT is becoming a vital instrument of interconnected devices for different applications, resulting in huge transformations of lives beyond imagination. It can result in optimization of resource usages and improved productivity in industry, becoming an essential part of life in distribution, retail management, transportation, healthcare services, farming, water resources, education, research, energy management (petroleum/atomic/solar), and administrative services. The IoT has great market-capturing potential, as all big names in the telecom and manufacturing industries and in government [65] are investing for IoT implementations.

The IoT has been identified as the most trending search on Google, as shown in search trends from 2014 to 2018. The popularity of wireless sensor networks seems to be consistent with time, but the in case of cloud computing, it has declined after 2014, whereas popularity of the IoT is soaring high. The comparative graph of the Google search trends for wireless sensor networks, IoT, an cloud computing clearly reflects that IoT is most trending area of research and it has a huge scope in the future. As per Google search forecasts, these trends are likely to continue, enabling more device integration for the new research areas that are yet to be invented (Figure 8.3).

The Internet of Things will bring new solutions for daily life problems, with wearable stuff that can be easily used and carried anywhere. These products must certainly impact our lives, making us capable of handling multiple devices simultaneously. The IoT is still in the phase of devising some standards for different IoT applications; the IoT has come a long way in terms of communication to connect different devices, providing ease of connecting and use of the devices. Through IoT we are able to keep a close eye on the area beyond our reach; like in a factory, we can see in real time what is going on in the factory, and the status of fuel and other important

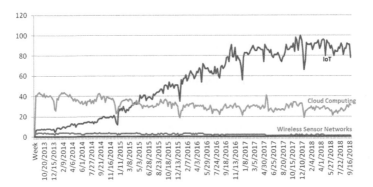

FIGURE 8.3
Trends of Google search, IoT, cloud computing, wireless sensor networks, for the last 5 years.

inventories. It's only a matter of time until all the devices will be connected through the Internet and can be controlled by mobile applications, leading to machine-to-machine communication and interfaces for talking to each other. We see a change in the lifestyle of millennial youth fond of wearable devices that track their daily activity and analyze the activity chart to know how much energy needs to be consumed for keeping them fit and healthy. These wearable devices are used for sugar-level tracking, BP level, and other vitals of the user. The IoT has helped a lot in keeping observation on environment, pollution, disease prevention, and traffic management.

IoT predictions by different sources are very promising, according to Lee [65] and Sundmaeker [66]; after 2020 cognitive networks, capable of self-repairing and self-learning, will come into existence. User-oriented software will be designed having features like being easy to deploy, and Things to Human collaboration will provide the IoT to all. Hardware features may converge nanotechnology with newly invented materials. Data processing will be optimized cognitive processing.

IoT can bring the following changes in the lives of people in society:

- **Smart Cities**: The smart city makes enormous use of the IoT. Smart cities are supposed to keep track of all resources in real time, be it water, air, intelligent transport systems [67], traffic management by installing GPS systems in vehicles [68], policing, domestic power and water billing, encroachment mapping and tracking, structural fitness of buildings [69], environment monitoring [70], critical infrastructure monitoring, solar power harvesting, water bodies observation, and greenery observation. In fruit and vegetable markets, IoT can help in tracking the condition of stored fruits and vegetables for contamination and hazards, if any, before consumption. Police, fire service, emergency service, and ambulance services can use IoT-based devices for finding the shortest routes to reach a destination in time to help the needy, as every second is important in an emergency. Smart light systems can be used in smart cities in which localized sensor nodes are connected to control the streetlights [71].

- **Smart Grid System**: The IoT has the potential to play an important role of sensing and monitoring the efficiency of current flow in the nodes of electricity networks. The IoT can help in providing information related to power consumption and quality of power supply to the farthest consumer node on the network. Some decision criteria is needed to find the best and most efficient route to transfer the power from the grid [72]. The IoT can improve the electrical grid system by sensing and analyzing real-time load for distribution network data and taking precautionary measures to avoid grid failure, ensuring flawless supply with quality energy for the high-priority consumers like hospitals, telecom industries, and water supply system. It can take care of the grid in peak load hours.

- **Industrial IoT (IIoT)**: The role of IoT is getting more prominence in smart manufacturing and industrial applications. The implementation of the IIoT [73] is going to completely change the scenario of manufacturing by connecting almost all applications through one control system called SCADA (supervisory control and data acquisition) [74]; this system collects real-time data and analyzes it, to take action as needed. The convergence of microsensors, robotics, cloud services, application software, data analytics, and mechanical processes has had a long-lasting impact on the manufacturing industry. The industrial IoT is also converging conventional IT (i.e., computers and machines) with OT (operational technology, i.e., control system hardware, sensors, actuators). The IIoT can be implemented in an environment unfriendly to the workers, like in high-temperature areas or poisonous fumes in paint jobs or other hazardous environments. Industry4.0 (I4.0) can be implemented through IIoT; I4.0 will be a reality soon, even before it is expected.

- **Smart Healthcare**: The IoT is already playing a major role in health monitoring [75,76] through wearable devices [77] that gather the real time data of the user. This data is stored on cloud and analyzed to get information about the health statistics of the user. There are a number of wearable non-incursive devices that are in an initial testing phase. These devices may collect the data related to all vital information (e.g., BP, blood sugar, oxygen level) [78]. This data is maintained on the cloud, creating patient history for forever. Hospitals are also being equipped with multiple connected devices giving information on a single display through an attachable probe. Medicines are administered as per the dosages required. Many IoT products that are still is labs, when installed may be able to monitor hundreds of patients with efficiency and save precious lives. E-health recommender [79] system is trending in research; in this, the patient is recommended for specialized facilities related to health services.

- **Agriculture and farming**: According to Christopher Brewster, the IoT is going to impact produce farming through different pilots [80] in Europe: dairy pilot, fruit pilot, arable crop pilot, and meat and vegetable pilots targeted for monitoring and controlling different parameters of their growth and maximum yield. The IoT can be used for avoiding misuse of water, fertilizers, and pesticides, resulting in cost cutting and good-quality produce [81]. The IoT can monitor [82] storage conditions. Innovations in the IoT are going on for processing, distribution, storage, and consumption analysis of food. Different devices based on IoT give real-time information on the health of farm animals, resulting in timely treatment and control on spreading of any communicable disease.

- **Service Industry**: The service industry [83] is going to change in a big way with the integration of IoT and service processes. Anyone can log on to the Internet and put in a service request: maybe for a hotel room, food, maintenance, water, building management, or environment monitoring. The IoT has a role to play—in water quality monitoring, leakage, and wastage monitoring and supply; in building maintenance, such as activity monitoring, temperature and humidity control, and HVAC (heating, ventilation, and air conditioning); and in the environment, such as pollution monitoring, water pollution in river monitoring, and industry pollution monitoring.

- **Defense and Security**: The IoT is massively used in defense applications, be it unmanned aerial vehicles [84], landmine sensors, thermal sensors for intrusion detection, or through-the-wall sensors to detect the presence of enemy behind the wall in a room. The IoT is also used to attack the enemy precisely through location services using the Internet, for visualization and identification.

References

1. https://www.internetworldstats.com/stats.htm.
2. http://www.faz.net/aktuell/wirtschaft/diginomics/grosse-internationale-allianz-gegen-cyber-attacken-15451953-p2.html?printPagedArticle=true#pageIndex_1.
3. G. Santucci, "The Internet of Things: Between the revolution of the Internet and the metamorphosis of objects" (PDF), *European Commission Community Research and Development Information Service*. Retrieved 23 October 2016.
4. IoT: Trends challenges and future. csjournals.com/IJCSC/PDF7-1/28.%20 Vyas.pdf.

5. J. Gubbi, R. Buyya et al., "Internet of Things (IoT): A vision, architectural elements, and security issues," *Future Generation Computer Systems* 29, (7): 1645–1660 (2013). https://www.elsevier.com/locate/fgcs.

6. https://ahmedbanafa.blogspot.com/2016/02/iot-implementation-and-challenges.html.

7. M. Thiyagarajan, C. Raveendra, "The framework of Internet of Things services," *International Conference on Computing, Communication and Automation(ICCCA2017)* 1282–1286 (2017).

8. List of temperature sensors, https://en.wikipedia.org/wiki/List_of_temperature_sensors. Last modified November 25, 2018.

9. https://electronicsforu.com/resources/electronics-components/humidity-sensor-basic-usage-parameter.

10. https://www.engineersgarage.com/articles/pressure-sensors-types-working.

11. https://www.azosensors.com/article.aspx?ArticleID=339.

12. http://thephotographerblog.com/definition-image-sensor/.

13. https://www.elprocus.com/optical-sensors-types-basics-and-applications/.

14. https://www.maximintegrated.com/en/app-notes/index.mvp/id/5830.

15. http://www.instrumentationtoday.com/accelerometer/2011/08/.

16. https://www.elprocus.com/working-of-different-types-of-motion-sensors/.

17. https://www.efxkits.co.uk/liquid-level-sensor-and-types-of-level-sensors/.

18. https://www.nfpa.org/Public-Education/By.../Smoke-alarms/Ionization-vs-photoelectric.

19. https://www.figaro.co.jp/en/technicalinfo/principle/electrochemical-type.html.

20. https://www.springer.com/in/book/9780387699301.

21. http://www.uky.edu/WDST/PDFs/[21.2]%20Water%20Quality%20Sensors.pdf.

22. https://www.ia.omron.com/support/guide/41/introduction.html.

23. S. S. I. Samuel, "A review of connectivity challenges in IoT-smart home," *Big Data and Smart City (ICBDSC), 2016 3rd MEC International Conference. IEEE,* (2016).

24. L. Li, H. Xiaoguang, C. Ke, H. Ketai, "The applications of WiFi-based wireless sensor network in Internet of Things and smart grid," 2011 *6th IEEE Conference on Industrial Electronics and Applications,* (2011).

25. K. Pahlavan, P. Krishnamurthy, A. Hatami, M. Ylianttila, J. P. Makela, R. Pichna, J. Vallstron, "Handoff in hybrid mobile data networks," *Mobile and Wireless Communication Summit* 7, 43–47 (2007).

26. X. Chen, L. Chen, M. Zeng, "Downlink resource allocation for device-to-device communication underlaying cellular networks," *2012 IEEE 23rd International Symposium on Personal, Indoor and Mobile Radio Communications—(PIMRC),* (2012).

27. X. Y. Chen, Z. G. Jin, "Research on key technology and applications for the Internet of Things," *Physics Procedia* 33, (2012): 561–566. doi:10.1016/j.phpro.2012.05.104.

28. P. Radmand, M. Domingo, J. Singh, J. Arnedo, A. Talevsk, S. Petersen, S. Carlsen, "ZigBee/ZigBee PRO security assessment based on compromised cryptographic keys," *2010 International Conference on P2P, Parallel, Grid, Cloud and Internet Computing,* (2010).

29. M. Franceschinis, C. Pastrone, M. A. Spirito, "On the performance of ZigBee Pro and ZigBee IP in IEEE 802.15.4 networks," *2013 IEEE 9th International Conference on Wireless and Mobile Computing, Networking and Communications (WiMob),* (2013).

30. S. Dong-feng, C. Xiang-jian, L. Di, "Research of new wireless sensor network protocol: ZigBee RF4CE," *2010 International Conference on Electrical and Control Engineering*, (2010).

31. M. B. Yassein, W. Mardini, A. Khalil, "Smart homes automation using Z-wave protocol," *2016 International Conference on Engineering & MIS (ICEMIS)*, (2016).

32. W. Twayej, H. S. Al-Raweshidy, "An energy efficient M2M routing protocol for IoT based on 6LoWPAN with a smart sleep mode," *2017 Computing Conference*, (2017).

33. D. Dragomir et al., "A survey on secure communication protocols for IoT systems," *International Workshop on Secure Internet of Things*, (2016).

34. W. Xue-fen, D. Xing-jing, B. Wen-qiang, L. Le-han, Z. Jian, Z. Chang, Z. Ling-xuan, Y. P. Yu-xiao, Y. I. Yang, "Smartphone accessible agriculture IoT node based on NFC and BLE," *2017 IEEE International Symposium on Consumer Electronics (ISCE)*, 78–79 (2017).

35. P. Vagdevi, D. Nagaraj, G. V. Prasad, "Home: IOT based home automation using NFC" *2017 International Conference on I-SMAC (IoT in Social, Mobile, Analytics and Cloud) (I-SMAC)*, (2017): 861–865.

36. M. Lauridsen, H. Nguyen, B. Vejlgaard, I. Z. Kovacs, P. Mogensen, M. Sorensen, "Coverage comparison of GPRS, NB-IoT, LoRa, and SigFox in a 7800 km² area," *2017 IEEE 85th Vehicular Technology Conference (VTC Spring)*, 2017: 1–5.

37. https://hackernoon.com/9-important-iot-protocols-a-developer-should-know-8541d0af9670.

38. W. Ayoub, M. Mroue, F. Nouvel, A. E. Samhat, J. C. Prévotet, "Towards IP over LPWANs technologies: LoRaWAN, DASH7, NB-IoT," *2018 Sixth International Conference on Digital Information, Networking, and Wireless Communications (DINWC)*, 2018: 43–47.

39. I. Z. Kovács, P. Mogensen, M. Lauridsen, T. Jacobsen, K. Bakowski, P. Larsen, N. Mangalvedhe, R. Ratasuk, "LTE IoT link budget and coverage performance in practical deployments," *2017 IEEE 28th Annual International Symposium on Personal, Indoor, and Mobile Radio Communications (PIMRC)*, 2017: 1–6.

40. http://www.rfwireless-world.com/IoT/.

41. H. Geng, "Networking protocols and standards for Internet of Things" in *Internet of Things and Data Analytics Handbook*, Hoboken, NJ: John Wiley & Sons, 2017: 816.

42. Radio-frequency identification, https://en.wikipedia.org/wiki/Radio-frequency_identification. Last modified December 15, 2018.

43. https://www.ecma-international.org/publications/files/ECMA-ST/ECMA-352.pdf.

44. J. Frahim, C. Pignataro, J. Apcar, M. Morrow, "Securing the Internet of Things: A proposed framework," Cisco White Paper, 015[online]. http://www.cisco.com/c/en/us/about/security-center/secure-IoT-proposed-framework.html#2.

45. M. Wolf, D. Serpanos, "Safety and security in cyber–physical systems and Internet-of-Things systems" *Proceedings of the IEEE* 106, (1): (2018).

46. J. Dalbey, "Nonfunctional requirements." Csc.calpoly.edu. Retrieved 3 October 2017.

47. C. Kalloniatis, E. Kavakli, S. Gritzalis, "Security requirements engineering for e-Government applications: Analysis of current frameworks," *International Conference on Electronic Government*. Berlin, Germany: Springer. (2004). https://link.springer.com/chapter/10.1007/978-3-540-30078-6_11.

48. J. D. Moffett, A. B. Nuseibeh, *A Framework for Security Requirements Engineering*. Report YCS 368, Department of Computer Science, York, UK: University of York. (2003).
49. E. Kavakli, S. Gritzalis, "Protecting privacy in system design: The electronic voting case," *Transforming Government: People, Process and Policy* 1, 307–332, (2007). (Online) www.emeraldinsight.com/1750-6166.htm.
50. M. Blaze, J. Feigenbaum, J. Lacy, "Decentralized trust management" in *Proceedings of IEEE International Symposium Security and Privacy*, Colorado Springs, pp. 164–173, (1996).
51. A. Ghosh, S. K. Das, "Coverage and connectivity issues in wireless sensor networks: A survey," *Pervasive and Mobile Computing* 4: 303–334 (2008).
52. S. S. I. Samuel, "A review of connectivity challenges in IoT smart home," *2016 3rd MEc International Conference on Big Data and Smart City*, (2016).
53. Todd Greene, "5 Challenges of internet of things connectivity," (2014) (online) https://www.pubnub.com/blog/2014-06-17-5-challenges-of-internet-of-things-connectivity/.
54. V. R. Konduru, M. R. Bharamagoudra, "Challenges and solutions of interoperability on IoT: How far have we come in resolving the IoT interoperability issues," *2017 International Conference on Smart Technologies for Smart Nation (SmartTechCon)* (2017).
55. A. P. Athreya, B. DeBruhl, "Designing for self-configuration and self-adaptation in the Internet of Things," *9th IEEE International Conference on Collaborative Computing: Networking, Applications and Work Sharing* (2013).
56. Z. Huang et al., "Communication energy aware sensor selection in IoT systems," *Proc. IEEE Int'l. Conf. Internet of Things: Green Computing and Communications and Cyber Physical and Social Computing*, 235–242 (2014).
57. M. D. Gutierrez, V. Tenentes, "Low cost error monitoring for improved maintainability of IoT applications," *2017 IEEE International Symposium on Defect and Fault Tolerance in VLSI and Nanotechnology Systems (DFT)* (2017).
58. M. Al-Zihad, S. A. Akash, T. Adhikary, "Bandwidth allocation and computation offloading for service specific IoT edge devices," *2017 IEEE Region 10 Humanitarian Technology Conference (R10-HTC)* (2017).
59. D. A. Vyas, D. Bhatt, D. Jha, "IoT: Trends, challenges and future scope," *International Journal of Computer Science & Communication* 7(1): 186–197 (2015). (online) csjournalss.com.
60. P. Semasinghe, S. Maghsudi, "Game theoretic mechanisms for resource management in massive wireless IoT systems," *IEEE Communications Magazine* 55, 2, (2017).
61. P. R. R. Papalkar, C.A. Dhote, "Issues of concern in storage system of IoT based big datax," *2017 International Conference on Information, Communication, Instrumentation and Control (ICICIC)* (2017).
62. V. Sharma, A. Vithalkar, M. Hashmi, "Lightweight security protocol for chipless RFID in Internet of Things (IoT) applications," *2018 10th International Conference on Communication Systems & Networks (COMSNETS)* (2018).
63. S. M. Kerner, "Hacking RFID tags is easier than you think: Black hat," (online) http://www.eweek.com/security/hacking-rfid-tags-is-easier-than-you-think-black-hat.

64. F. Khan, "Future scope and possibilities in internet of things," *AESM, International Conference on Advances in Engineering Science and Management.*

65. K. Lee, "Internet of things (IoT): Applications, investments and challenges for enterprises," published by Elsevier (online) https://www.science direct.com.

66. H. Sundmaeker, P. Guillemin, P. Friess, S. Woelffle (2010), Vision and challenges for realising the Internet of Things. Accessible at http://www.researchgate.net/publication/228664767_Vision_and_challenges_for_realising_the_Internet_of_Things.

67. K. Jakobs, "Standardisation of e-merging IoT-applications: Past, present and a glimpse into the future," *2016, 4th International Conference on Future Internet of Things and Cloud Applications. IEEE Computer Society* (2016).

68. X. Li, W. Shu, M. Li, H. Y. Huang, P. E. Luo, M. Y. Wu, "Performance evaluation of vehicle-based mobile sensor networks for traffic monitoring," *IEEE Trans. Veh. Technol.,* 58(4): 1647–1653 (2009).

69. J. P. Lynch, J. L. Kenneth, "A summary review of wireless sensors and sensor networks for structural health monitoring," *Shock and Vibration Digest,* 38(2): 91–130 (2006).

70. N. Maisonneuve, M. Stevens, M. E. Niessen, P. Hanappe, L. Steels, "Citizen noise pollution monitoring," in *Proceeding 10th Annual. International. Confernce. Digital Government. Research: Social Network: Making Connection Between Citizens, Data Government,* 96–103 (2009).

71. A. Zanella, N. Bui, A. Castellani, L. Vangelista, M. Zorzi, "Internet of Things for smart city," *Ieee Internet of Things Journal* 1, 1 (2014).

72. R. Morello, C. De Capua, G. Fulco, "A smart power meter to monitor energy flow in smart grids: The role of advanced sensing and IoT in the electric grid of the future," *IEEE Sensors Journal* 17, 23, (2017).

73. H. Sasajima, T. Ishikuma, H. Hayashi, "Future IIOT in process automation—Latest trends of standardization in industrial automation, IEC/TC65," *Society of Instrument and Control Engineers of Japan Annual Conference* July 28–30, 2015, Hangzhou, China (2015).

74. R. Hunzinger, "Scada fundamentals and applications in the IoT," in *Internet of Things and Data Analytics Handbook,* First Edition. Edited by Hwaiyu Geng. © 2017 Hoboken, NJ: John Wiley & Sons. (2017). Companion website: https://www.wiley.com/go/Geng/iot_data_analytics_handbook/.

75. S. V. Zanjal and G. R. Talmale, "Medicine reminder and monitoring system for secure health using IOT," *Physics Procedia,* 78, no. December 2015, 471–476, (2016).

76. G. S. Tamizharasi, H. P. Sultanah, "IoT-based e-health system security: A vision architecture elements and future directions," *International Conference on Electronics, Communication and Aerospace Technology ICECA* (2017).

77. S. Sonune, D. Kalbande, "IoT enabled API for secure transfer of medical data," *2017 International Conference on Intelligent Computing and Control (I2C2).* (2017).

78. S. K. Datta et al., "Applying Internet of Things for personalized healthcare in smart homes," *Wireless and Optical Communication Conference (WOCC),* 2015 24th. IEEE (2015).

79. H. J. Lee, H. S. Kim, "eHealth recommendation service system using ontology and case-based reasoning," *2015 IEEE International Conference on Smart City/SocialCom/SustainCom together with DataCom 2015 and SC2 2015* (2015).

80. C. Brewster, I. Roussaki, N. Kalatzis, K. Doolin, K. Ellis, "IoT in agriculture: Designing a Europe-wide large-scale pilot," *IEEE Communications Magazine* 55 (9): 26–33 (2017).
81. R. Venkatesan, A. Tamilvanan, "A sustainable agricultural system using IoT," *International Conference on Communication and Signal Processing*, April 6–8, 2017, India.
82. H. Jun-Wei, Y. Shouyi, Z. Zhen, "A crop monitoring system based on wireless sensor network," *Procedia Environmental Sciences* 11, 558–565 (2011).
83. S. Meyer, A. Rupeen, C. Magerkurt. "Internet of Things—Aware process modeling: Integrating IoT devices as business process resources," *International Conference on Advanced Information Systems Engineering.* Berlin, Germany: Springer, 84–98 (2013).
84. P. Fraga-Lamas, T. M. Fernández-Caramés, M. Suárez-Albela, L. Castedo, M. González-López, *A Review on Internet of Things for Defense and Public Safety* (online). https://www.mdpi.com/sensors.
85. D. D. Sanju, A. Subramani, V. K. Solanki, "Smart city: IoT based prototype for parking monitoring & parking management system commanded by mobile app," in *Second International Conference on Research in Intelligent and Computing in Engineering* (2017).
86. R. Dhall, V. K. Solanki, "An IoT based predictive connected car maintenance approach" *International Journal of Interactive Multimedia and Artificial Intelligence* (ISSN 1989-1660), 4(3): 2017.
87. V. K. Solanki, M. Venkatesan, S. Katiyar, "Conceptual model for smart cities for irrigation and highway lamps using IoT," *International Journal of Interactive Multimedia and Artificial Intelligence*, (ISSN 1989-1660), 1, (2018).
88. Vijender Kumar Solanki, M. Venkatesan, S. Katiyar, "Think home: A smart home as digital ecosystem in circuits and systems," *Scientific Research Publishing* 10, 7, (2018).
89. V. K. Solanki, S. Katiyar, V. Bhaskar Semwal, P. Dewan, M. Venkatesan, N. Dey, "Advance automated module for smart and secure city," in *ICISP-15*, organised by G.H. Raisoni College of Engineering & Information Technology, Nagpur, on 11–12 December 2015, published by Procedia Computer Science, Amsterdam, the Netherlands: Elsevier.
90. V. G. Kadam, S. C. Tamane, and V. K. Solanki, "Smart and connected cities through technologies" in *Big Data Analytics for Smart and Connected Cities*, Hershey, PA:IGI Global, 2019: 1–24.

9

Learner to Advanced: Big Data Journey

Meenu Gupta and Neha Singla

CONTENTS

9.1 Introduction

The idea of massive knowledge in data technology pertains to a sequence of approaches, tools, and strategies for handling structured and unstructured knowledge, and enormous amounts of major diversity for the results. It is acknowledged by people to be effectual in conditions of uninterrupted growth and distribution across multiple nodes of an electronic network. The conception was fashioned at the top of the 2000s, different from the old direction systems and business intelligence category solutions. Massive knowledge may be an idea of covering many aspects by one term, starting from a knowledge base to a group of monetary models. A definition [1] of "big data" is as follows:

> *By extending it we can say that the platform, tools and software used for this purpose are collectively called big data technologies.*

Examples for supply of huge knowledge square measure include knowledge from activity devices, weather-related facts, message streams from social networks, incessantly incoming events of frequency identifiers, knowledge streams on the situation of subscribers of cellular networks devices, video soundtracks, remote sensing, and audio recordings. Huge knowledge isn't associated with nursing unexplained ideas, but rather with moving targets coupled to a technology context (Figure 9.1).

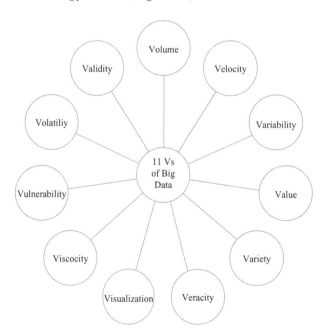

FIGURE 9.1
Big data.

People who work in academia and industry, as well as other eminent stakeholders, positively accord that big data has became a big game-changer in most contemporary industries. There has been a significant shift of focus, from neighboring on the hard sell to instead finding real value in its use as big data continues to permeate of our day-to-day lives.

Understanding big data values is a big challenge, but funding, return investment and skills are other big challenges for those industries who adopt big data. A survey from 2015 [2] concludes that more than 75% of companies were interested in investing or planning to invest in big data in upcoming years. These outcomes represent a substantial increase from an analysis done in 2012 [3], which indicated that 58% companies were invested or planning to invest in big data in the coming years.

9.2 Motivation of This Chapter

The main focus of this chapter is to enable the learner to advance in a journey into big data. The concept of big data exists from many years ago, but now the majority of associations emphasize that if they capture all the data, they can stream it into their business and apply analytics to get significant value from it. Even in the era of the 1950s, people were not aware of the term "big data"; they were using spreadsheets and manual examinations to uncover insights and trends.

Long ago, organizations were collecting information about data, running analytics on it, and unearthing information. Furthermore, it could be used for future decision, in that businesses could identify perceptions for immediate decisions in everyday life. It gave organizations a modest edge, the ability to work faster and stay agile, that they didn't have before. In the case of advanced big data analytics, machine learning (ML) algorithms are increasingly being used in order to automate the analytics process and also for the better management of the three V's of big data: volume, velocity and variety. It is the ongoing explosion in the availability of large and complex business datasets.

9.3 Comparison among Business Intelligence, Analytics, and Big Data

9.3.1 Business Intelligence

This helps many organizations to make better decisions by providing a wide range of the latest tools and methods. It also includes varied processes and procedures for solicitation of data. It involves sharing and reporting in order to ensure better decisions. With the latest advances in business intelligence (BI) tools, users can generate stories and visualizations all by themselves, without relying on other technical staff.

According to [4] Mark van Rijmenam, the CEO and founder of Big Data Startups, "the difference between Business Intelligence and Data Analytics dwells the fact that based on past results, Business Intelligence supports in making business decisions while data analytics facilitates in making predictions that helps you in the future."

Conventionally, we can say that BI deals with analytics and reporting tools that helps in influencing trends that are based on historical data. According to [5] Pat Roche, vice president of engineering for Noetix Products' magnitude software, "Business Intelligence is used to operate the business whereas Business Analytics is used to transform the business."

9.3.2 Business Analytics

The increase in value for incorporation as well as association of information is shown by the latest trends in business analytics in order to certify policy formation and encounter strategic purposes. In order to meet ever-growing data technology needs with extended capabilities, several companies are utilizing hi-tech business tools. A recent survey [6] in *Forbes* magazine said, "The Big Data trends led to almost 15% growth in this arena last year alone which affects the data analysis in different ways."

9.3.3 Data Science

Data science is a recent field of combines big data, unstructured data and also advanced mathematics and statistics. It is a new field, providing understanding of correlations between structured and unstructured data that has emerged within the field of data management. Nowadays, the demand of data scientists has been increased, as they can transform unstructured data into actionable understandings that are helpful for businesses.

9.4 Big Data with Hadoop

The term "big data" is used to account large datasets of different items. Data storage, management, search, analytics, and alternative process became a challenge. It is described by the degree of digital information that may come from several sources and information formats such as structured as well as unstructured, in order to find perspective and structure that are used to make informed decisions based on the data that can be analyzed and processed. By streamlining the execution of data-intensive, highly parallel distributed applications, the encounters of big data can be met by Apache Hadoop.

For stowing huge quantities of facts, Hadoop provides an economical manner. It facilitates a scalable and reliable mechanism, over a collection of product of similar type, for processing large volumes of data. Also, it offers original and enhanced analysis techniques that allow complicated systematic processing of multiple structured data. Hadoop is dissimilar from earlier distributed approaches in the following ways:

- Spreading of facts is in advance.
- For reliability and availability the data is simulated throughout a cluster of computers.
- Data processing can perform where the data is stored to remove bandwidth chokepoints.

9.4.1 Principal Components of Hadoop Ecosystem

Following are the main components of the Hadoop ecosystem:

- **Hadoop Distributed File System (HDFS):** It is a primary element of the Hadoop ecosystem. It's a method by which an outsized volume of information can be distributed over a cluster of computers. During this, information will write once and read multiple times for analysis. This also acts as a creator for supplementary tools, such as HBase.
- **MapReduce:** MapReduce is an important execution structure of Hadoop that could be a programming model for distributed, parallel processing by splitting jobs into two different segments such as the mapping stage and reduce stage. By exploiting information kept in HDFS for quick information access, MapReduce jobs for Hadoop are written by developers.
- **HBase:** A column-oriented NoSQL database built on top of HDFS. It is used for quick read/write access to a big volume of data. HBase uses Zookeeper for its management to make sure that every one of its parts is up and running.
- **Zookeeper:** The Hadoop's distributed coordination service is understood as Zookeeper. It's designed to run over a cluster of machines. It's an extremely accessible service used for the management of Hadoop operations. Also, the varied elements of Hadoop depend upon it.
- **Oozie:** Oozie is incorporated into the Hadoop stack. It is a scalable workflow system that's customized to coordinate execution of assorted MapReduce jobs. It does basically depend upon implementation of peripheral events that embody temporal order and presence of needed knowledge; it's capable of managing a major degree of complexity.

- **Pig:** An abstraction over the complexity of MapReduce programming, the Pig platform is associated with nursing execution surroundings and a scripting language (Pig Latin) wont to analyze Hadoop information sets. The compiler of Pig translates Pig Latin into sequences of MapReduce programs.

- **Hive:** Hive was developed on abstraction layer to gear a lot of towards database analysts that is lot of well-known with SQL rather than Java programming. It is an SQL-like, application-oriented language customized to run queries on information keep in Hadoop. In order to write information queries that are translated into MapReduce jobs in Hadoop, Hive allows developers who don't seem to be acquainted with MapReduce.

9.4.2 Hadoop Architecture

It is a free source software structure. It is used for storing as well as big-scale processing of information sets that lie on clusters of commodity hardware. This runtime environment consists of mainly five building blocks (from bottom to top) (Figure 9.2).

The ordinary file system used by Hadoop is HDFS, which uses master/slave structural design. The master consists of a one NameNode and more than one slave DataNodes within which the NameNode manages the file system information and also the actual facts stored by DataNodes.

In an HDFS namespace, a document is separated into a number of blocks and a further set of DataNodes are customized to store those blocks.

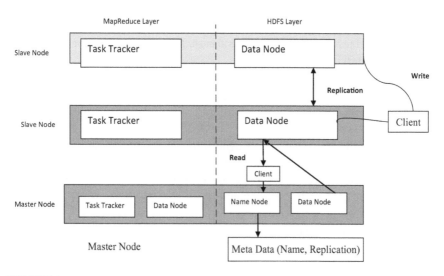

FIGURE 9.2
Hadoop architecture.

The NameNode regulates the mapping of blocks to the DataNodes. The read and write operations within the file system are determined by DataNodes, and also block creation, deletion and replication supported instruction given by NameNode are determined by it.

9.5 Advanced MapReduce

Apache Hadoop is an open resource computing code package scaffold. It can be run on a cluster of product machinery. So as to store and method massive amounts of knowledge during an extremely distributed manner, the machines will communicate and work together. From the start, there have been two main elements of Hadoop: a Hadoop Distributed file system (HDFS) and, therefore, the second may be a distributed computing engine. It permits you to employ and execute programs as MapReduce jobs.

One of the foremost necessary parts of Hadoop is the infrastructure. It's answerable for all the composite aspects of distributed process that have hardware failures, parallelization, scheduling, resource management, inter-machine communication, covering software package, and so on. Implementing distributed applications that use terabytes of information on a whole bunch (or even thousands) of machines is the main advantage of fresh abstractions. It was not really as easy and conjoined as now, even for developers who do not have any not have any previous experience with distributed systems.

9.5.1 Limitations of Classical MapReduce

The conventional MapReduce is fundamentally associated with measurability and utilization of resources; and conjointly, the hold-up of workloads that are dissimilar from MapReduce is one of its most serious limitation. Within the MapReduce structure, the task execution is restricted by two kinds of processes:

- A JobTracker that is a solitary master process: On the cluster, it coordinates all jobs that are running. After that it notates map and reduce tasks to execute on the TaskTrackers.
- A TaskTrackers is a variety of subordinate processes. It runs appointed tasks; however, it sporadically reports to achieve the JobTracker (Figure 9.3).

Having a single JobTracker could be a reason for the outsized Hadoop clusters that expose a restriction, including a scalability bottleneck. According to [7] Yahoo!, with a cluster of five thousand nodes and forty thousand tasks that are running at the same time, the practical limits of such a design are

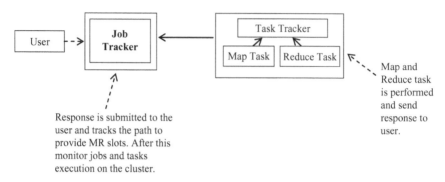

FIGURE 9.3
Classical version of Apache Hadoop (MRv1).

reached, and slighter and less powerful clusters have to be shaped and maintained because of this limitation.

Initially, Hadoop would frame to execute MapReduce tasks only. With the advancement of different programming models, there was a huge increase required to sustain programming paradigms as well as those of MapReduce. It shall be executed on the identical cluster and also share resources in an efficient and unbiased manner.

In 2010, engineers at Yahoo! began functioning on a totally innovative design of Hadoop [8]. It addresses all the restrictions that square measure represented on top of multiple extra options conjointly.

9.6 Hadoop 2.0, YARN

For big data processing, Apache Hadoop is one in every of the foremost widespread tools. It has been effectively executed in production by several firms for many years. It's continuously being improved by an over-sized community of developers. Even supposing Hadoop has taken into account a reliable, scalable, and efficient answer that ends up in many revolutionary options, Yet Another Resource Negotiator (YARN), HDFS Federation, and an extremely offered Name Node offered by version 2.0 together make the Hadoop cluster much more efficient, powerful, and reliable.

By slightly changing a language, the subsequent name changes offer a morsel of eminence into the planning of YARN:

- Resource Manager rather than cluster manager
- Application Master rather than a dedicated and short-lived JobTracker

- Node Manager rather than Task Tracker
- A distributed application rather than a MapReduce job

9.6.1 Architecture of YARN

YARN architecture is a dedicated machine of a global resource manager that runs as a master daemon and also arbitrates the obtainable cluster resources among numerous competitive applications. What percentage of live nodes and resources are available on the cluster are determined by the Resource Manager. It also takes a look about type of applications are submitted by users and what these resources acquire and at what time. A client can submit any type of application supported by YARN. The Resource Manager is the solitary process. This information helps to make its allocation (or rather, scheduling) decisions in a secure, shared and multi-tenant manner.

The more generic and efficient version of the Task Tracker is Node Manager. The Node Manager encompasses a variety of dynamically created resource containers rather than having a defined number of maps and reduces slots. The quantity of resources it contains, like memory, CPU, network IO, and disk are used to categorize the size of a container. At present, only memory and CPU (YARN-3) are in its support. The number of containers on a node is an artifact of configuration parameters and the whole amount of node resources that comprise total CPUs and overall memory that is not inside the resources devoted to the slave daemons and the operating system.

Inside a container, an Application Master could execute any kind of activities. For instance, the MapReduce Application Master wishes a container to begin a map or a reduce task, whereas the Giraph Application Master wishes a container to execute a Giraph undertaking. This could also execute a custom Application Master. It executes definite tasks. In this manner, it changes the big data world by inventing a shiny new distributed application framework. MapReduce is directly reduced to a position of a distributed application. YARN is currently additionally known as MRv2. MRv2 is just twice the implementation of the conventional MapReduce engine that is currently additionally referred to as MRv1. It runs on the high end of YARN. A container can run diverse types of tasks and have different sizes of RAM and CPU (Figure 9.4).

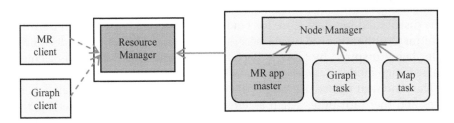

FIGURE 9.4
YARN architecture.

YARN is nothing but a thoroughly rewritten architecture of the Hadoop cluster. The way distributed applications are implemented, it seems to be a game-changer. It is also executed on a cluster of commodity machines.

9.7 Introduction of Other Apache Tools for Big Data and Machine Learning Tasks

9.7.1 Mahout

Mahout is known to be a popular data mining library. For performing clustering, classification, regression, and statistical modeling to prepare intelligent applications, Mahout takes the most popular data mining scalable machine learning algorithms. The main objective of Apache Mahout is to increase a responsive, vibrant and diverse community in order to facilitate discussions that are not only based on the project itself but also on potential use cases.

A list of the companies that use Mahout:

- **Amazon:** It is an online purchasing doorway for providing individual suggestion.
- **AOL:** This is a purchasing site for shopping recommendations.
- **Drupal:** It is a PHP content management system using Mahout for providing open source content-based recommendation.
- **Twitter:** It is a social networking website that uses Mahout's Latent Dirichlet Allocation (LDA) accomplishment for user interest modeling and maintains a fork of Mahout on GitHub.
- **Yahoo!:** This is the world's most well liked Internet service supplier that uses Mahout's Frequent Pattern Set Mining for Yahoo! Mail.

9.7.2 Apache HBase

The distributed big data store for Hadoop is Apache HBase, which permits arbitrary, authentic read/write access to big data. This is often designed, after being inspired by Google BigTable, as an innovated column-oriented data storage model.

The canonical HBase use case is the web table. It is a table of crawled websites and their attributes (such as language and MIME type) keyed by the online page uniform resource locator. The online table is very big. It consists of row counts that run into the billions.

The lists of the companies using HBase are defined below:

- **Yahoo!:** one of the foremost world's fashionable net service suppliers for close to replica document recognition.
- **Twitter:** This is a community networking web site for version management storage and retrieval.
- **Mahalo:** It's an information distribution service for related content reference.
- **NING:** This service contributes to public for actual time analytics and reporting.
- **Stumble Upon:** It's a worldwide adapted suggested system, real time data storage, and information analytics platform.
- **Veoh:** It is an internet multimedia system information sharing platform for consumer identification system (Figure 9.5).

9.7.3 Hive

In the information platform, Hive is known to be one of the biggest ingredients. Jeff Hammerbacher's team at Facebook built Hive, which is a framework for data warehousing on top of Hadoop. Hive grew from a need to manage and learn from the huge volumes of data. From its burgeoning social network, it was then further produced by Facebook every day. After trying a few different systems, Hadoop was chosen by the team for storage and processing, as it was effective in terms of cost and also met the scalability requirements.

Hive allows users to fire queries in SQL-like languages, such as HiveQL, which are highly abstracted to Hadoop MapReduce, which then further allows SQL programmers with no MapReduce experience to use the warehouse and makes it easier to integrate with business intelligence and visualization tools for real-time query processing (Figure 9.6).

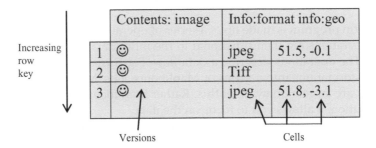

FIGURE 9.5
The HBase data model, instance for a table recording images.

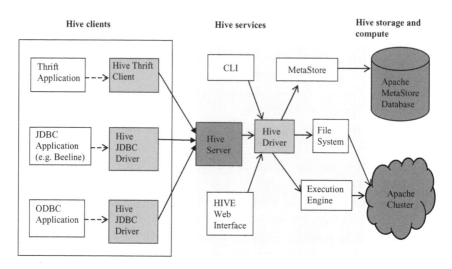

FIGURE 9.6
HIVE architecture.

9.7.4 Pig

For processing large datasets, Apache Pig raises the level of abstraction. As the programmer, MapReduce permits you to identify a map function, which can be followed by a diminish function, which is then working out how to fit your information processing into this prototype. It usually needs multiple MapReduce stages that may also be a challenge.

Pig is a group of two pieces:

1. Pig Latin is responsible for flow of data.
2. The execution is set so as to run programs of Pig Latin. There, square measure presents two environments: during a single JVM, local execution; and on a Hadoop cluster, distributed execution.

A Pig Latin program is made up of a series of operations, or transformations. These operations are then functional to the input data in order to generate output. The operations describe a data flow. Pig turns, under the blankets, the transformations into a series of MapReduce jobs. Merely as a programmer, you are largely unaware of this. Rather than focusing on the nature of the execution, it allows you to focus on the data.

In a few situations, Pig doesn't perform well in addition to programs written in MapReduce. Because the Pig team implements refined algorithms for applying Pig's relative operators, the gap is narrowing with each release. It's fair to say that writing queries in Pig Latin can lose time, unless you're willing to take a position of plenty of effort in optimizing Java MapReduce code.

For analyzing the big-scale datasets via its own SQL-like language, Pig Latin is a Hadoop-based open source platform. For massive, complex data-parallelization computation, it gives a straightforward operation and programming interface. It is also easier to enlarge and also, it's more optimized and extensible. Yahoo! developed Apache Pig. Currently the primary Pig users are Yahoo! and Twitter. For developers, Pig limits the Java programmer's use of Hadoop programming's flexibility, but the direct use of Java APIs can be tedious or error-prone. Hence in order to make Hadoop programming easier for dataset management and dataset analysis with MapReduce, Hadoop provides two solutions: these are Pig and Hive.

9.7.5 Apache Sqoop

By quickly transferring large amounts of data in a new way, Apache Sqoop provides Hadoop data processing platform and relational databases, data warehouse, and other non-relational databases. Apache Sqoop is a mutual data tool for importing data from the relational databases to Hadoop HDFS by exporting data from HDFS to relational databases. It works together with most modern relational databases, which include MySQL, PostgreSQL, Oracle, Microsoft SQL Server, and IBM DB2, and enterprise data warehouse. Scoop extension API provides a path to create new connections for the database system. Likewise, the Sqoop source comes up with some popular database connectors. With some logic of database schema creation and transformation, Sqoop first transforms the data into Hadoop MapReduce in order to perform this operation (Figure 9.7).

9.7.6 Apache ZooKeeper

Apache ZooKeeper is represented as a Hadoop sub project which is responsible for handling Pig, HBase, Hadoop, Hive, Solr, and other projects. ZooKeeper is an open source distributed applications coordination service. It is designed with Fast Paxos algorithm, which is based on synchronization, configuration and naming services such as upholding of distributed applications.

ZooKeeper is split into two parts: the server and consumer. For a cluster of ZooKeeper servers, it simply works as an organizer. The remainder of the

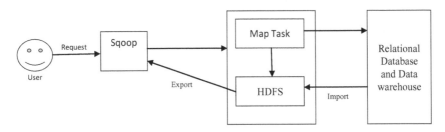

FIGURE 9.7
Sqoop's import process.

server is read-only copy of the master. Another server will begin allocating all requests, if the leader server goes down. ZooKeeper consumers are associated with a server on the ZooKeeper service. The consumers send an acknowledgement, get a response, access the viewer activities, and launch a heartbeat via a TCP correlation with the server.

Apache Solr is an open source organization looking for a base from the Apache license project. It's extremely ascendable, sustaining dispersed findings and index replication engine by permitting building Internet request for controlling text hunt, conditional hunt, authentic time period categorization, active cluster, information incorporation, and affluent document conduct. Because of these options, these findings are employed by Netflix, AOL, CNET, and Zappos.

The Prons of ZooKeeper can be characterize as following:

- **It is easy:** It is easy at its central part. It's an uncovered classification system that exposes many easy operations and furthermore includes a few additional abstractions, like ordering and notifications.

- **It is communicative:** A rich set of building blocks is a primitive of ZooKeeper, which may be utilized to construct a oversized category of coordination like knowledge structures and protocols. An example consists of distributed queues, distributed locks, and organizer determination among a bunch of equals.

- **ZooKeeper is extremely obtainable:** It executes on a group of machinery and is intended to be extremely useable, thus applications will rely on it. It will assist you to avoid introducing single points of breakdown into your organization in parliamentary procedure to make a real application.

- **It provides loosely coupled communications:** This communication support participants that ought not to understand one another. For instance, ZooKeeper may be used as an appointment mechanism in an order that processes. It doesn't grasp every other's existence (or network details), who would otherwise discover and move with each other. Since one method might drop a message in ZooKeeper that's scanned by another when a primary has been finished off by coordinative parties, it might not even be contemporaneous.

9.7.6.1 It Works as a Library

It works on open supply with shared depositories of execution and receives universal coordination patterns (Figure 9.8).

9.7.7 Ambari

Ambari is very specific to Hortonworks. Apache Ambari is a web-based tool that supports Apache Hadoop cluster supply, management, and

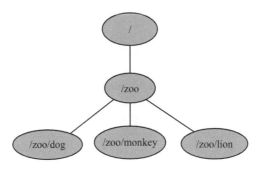

FIGURE 9.8
ZooKeeperznodes.

monitoring. It handles almost all of the Hadoop components, including HDFS, MapReduce, Hive, Pig, HBase, Zookeeper, Sqoop, and HCatlog as a centralized management. In addition to this, Ambari is able to install security based on the Kerberos authentication protocol over the Hadoop cluster. It also provides role-based user auditing functions, authentication, and authorization for users to manage integrated LDAP and Active Directory.

9.7.8 Spark

This technology was intended for quick computation, and this depends upon Hadoop MapReduce for more types of computations, by extending the MapReduce model in order to use it efficiently.

Spark is meant to handle a broad variety of workloads like repetition algorithms, interactive questions, batch applications and streaming. It minimizes the organization load of maintaining detach tools, excluding supporting of these workloads in an exceedingly individual system.

Features of Apache Spark are as follows:

- **Speed:** It is responsible to execute an application in Hadoop cluster. By reducing the variety of R/W operations to disk, it can be a hundred times quicker in memory and ten times quicker on disk. It lies within the intermediate processing data in storage.
- **Supports multiple languages:** It offers integrated application programming interface in Java, Scala and Python. This is used to comprise applications in multiple linguistic. Eighty high level operators are used by spark for interactive querying.
- **Advanced analytics:** "Map" and "reduce" task is performed by spark even though it refers SQL queries, Streaming data, Machine learning (ML), and Graph algorithms.

A Resilient Distributed Datasets is an unchallengeable scattered collection of objects where every data is split into logical portions for another execution on different nodes of clusters. It is a read only separated group of records.

A resilient distributed dataset (RDD) is split into two different ways: one is parallelizing the present group of data in drives, and another one is referencing the data in auxiliary memory that includes a shared file system, HDFS, HBase and other data proposed by other input formats of Hadoop. Spark uses RDD to get faster and more capable MapReduce operation.

9.7.8.1 Spark Built on Hadoop

Spark could be worked with a Hadoop component using three different methods; these three methods of Spark procedure are defined as follows:

- **Standalone:** A Spark standalone procedure suggest that Spark contains an area over the top of HDFS and that areas are allotted for HDFS unambiguously. At this point, Spark and MapReduce can run aspects by aspects to wrap all Spark jobs on the cluster.
- **Hadoop YARN:** Generally, the exploitation of Hadoop Yarn specifies that Spark could run on YARN with no pre-installation or any prerequisite of root access. This helps to put together Spark into the Hadoop ecosystem or Hadoop stack by permitting alternative parts to execute on the apex of stack.
- **Spark in MapReduce (SIMR):** It is a customized, open Spark task to additionally separate preparation. During this the user will begin Spark and use its shell with no body access.

9.7.9 Components Architecture of Spark

Various elements of Spark are specified in Figure 9.9.

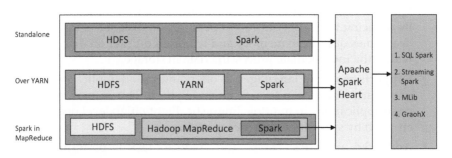

FIGURE 9.9
Component architecture of spark.

9.7.9.1 Apache Spark Heart

The heart of Spark is the fundamental all-purpose realization engine for the Spark platform where every alternative practicality is constructed upon. This is available in storage compute and referencing information in peripheral memory systems.

9.7.9.2 SQL Spark

Spark SQL may be an element on top of Spark heart that includes a novel data abstraction known as a design of RDD. It additionally facilitates the hold-up for structured and partially structured facts.

9.7.9.3 Streaming Spark

This supports Spark heart's quick forecast facility to execute streaming analytics. It ingests facts in small parts and performs RDD transformations on these small parts of facts.

9.7.9.4 ML lib (Machine Learning Library)

ML lib could be a dispersed machine learning structure over the peak of Spark as a result of the distributed memory-based Spark design. On the basis of the prototype, this could be completed by the ML lib developers in opposition to the alternating statistical method (ALS) execution. It's nine times as quick because of the Hadoop disk-based version of Apache driver (earlier than the driver gained a Spark edge).

9.7.9.5 GraphX

On top of Spark, GraphX could be a distributed graph-processing framework. For expressing graph computation, it provides an API, which will model the user-defined graphs that are then done by using Pregel abstraction application programming interface. For this abstraction, it additionally gives an optimum runtime environment.

References

1. Hikri, K. (2018). Understanding the four pillars of big data analytics. https://www.pluralsight.com/guides/understanding-the-four-pillars-of-big-data-analytics. Accessed December 18, 2018.
2. Tsai, C.-W., Lai, C.-F., Chao, H.-C., and Vasilakos, A. V. (2015). Big data analytics: A survey. *Journal of Big Data*, 2(21). https://journalofbigdata.springeropen.com/articles/10.1186/s40537-015-0030-3.

3. Davenport, T. H. (2012). Big data executive survey 2012. *NewVantage Partners*, pp. 1–23. Accessed: http://newvantage.com/wp-content/uploads/2012/12/NVP-Big-Data-Survey-2012-Consolidated-Report-Final1.pdf.
4. Williams, J. (2015). Business intelligence vs data analytics. https://www.promptcloud.com/blog/business-intelligence-Vs-data-analytics. Accessed July 28, 2015.
5. Roche, R. (2012). Executive Q & A with Pat Roche, Vice President Engineering Applications, Magnitude Software. https://magnitude.com/blog/qa-pat-roche-vp-engineering-apps/. Accessed March 2018.
6. Porres, E. L. (2013). The big potential of big data. https://images.forbes.com/forbesinsights/StudyPDFs/RocketFuel_BigData_REPORT.pdf. Accessed March 14, 2018.
7. Shvachko, K. V., and Murthy, A. C. (2008). Scaling Hadoop to 4000 nodes at Yahoo. http://yahoohadoop.tumblr.com/post/98117096126/scaling-hadoop-to-4000-nodes-at-yahoo. Accessed July 2018.
8. Ahrens, M. (2015). The history of large-scale data pipelines (Part 1): The migration to Hadoop. http://yahoohadoop.tumblr.com/page/3. Accessed September, 2018.

Further Reading

Apache Hadoop, accessed https://hadoop.apache.org/docs/r1.2.1/hdfs_design.html.

Chen, H., Chiang, R. H., and Storey, V. C. (2012). Business intelligence and analytics: From big data to big impact. *MIS Quarterly*, 1165–1188.

Dhall, R., and Solanki, V. K. (2017). An IoT based predictive connected car maintenance approach, *International Journal of Interactive Multimedia and Artificial Intelligence*, 4(3), 16–22.

Dr. Diane Schleier-Keller (2017). Top 3 reasons to invest in big data. https://inside bigdata.com/2017/08/23/top-3-reasons-invest-big-data/. Accessed October 27, 2017.

Franks, B. (2012). *Taming the Big Data Tidal Wave: Finding Opportunities in Huge Data Streams with Advanced Analytics* (Vol. 49). Hoboken, NJ: John Wiley & Sons.

Gollapudi, S. (2016). *Practical Machine Learning*. Birmingham, UK: Packt Publishing.

Google Cloud. https://cloud.google.com/solutions/big-data/.

Hadoop (2012). *The Definitive Guide*. O'Reilly Media. Foster, I., Ghani, R., Jarmin, R. S., Kreuter, F., and Lane, J. (2016). *Big Data and Social Science: A Practical Guide to Methods and Tools*. Boca Raton, FL: Chapman & Hall/CRC.

Holmes, A. (2012). *Hadoop in Practice*. Shelter Island, NY: Manning Publications.

Kadam, V., Tamane, S., and Solanki, V. Smart and connected cities through technologies, IGI-Global, doi:10.4018/978-1-5225-6207-8.

Kulkarni, A. P., and Khandewal, M. (2014). Survey on Hadoop and Introduction to YARN. *International Journal of Emerging Technology and Advanced Engineering*, 4(5), 82–87.

Mendelevitch, O., Stella, C., and Eadline, D. (2016). *Practical Data Science with Hadoop and Spark: Designing and Building Effective Analytics at Scale*. Boston, MA: Addison-Wesley Professional.

Moorthy, J., Lahiri, R., Biswas, N., Sanyal, D., Ranjan, J., Nanath, K., and Ghosh, P. (2015). Big data: Prospects and challenges. *Vikalpa*, 40(1), 74–96.

Murthy, A. C., Vavilapalli, V. K., Eadline, D., and Markham, J. (2014). *Apache Hadoop YARN: Moving Beyond MapReduce and Batch Processing with Apache Hadoop 2*. Upper Saddle River, NJ: Pearson Education.

Sanju, D. D., Subramani, A., and Solanki, V. K. (2017). Smart city: IoT based prototype for parking monitoring & parking management system commanded by mobile app, in *Second International Conference on Research in Intelligent and Computing in Engineering*.

Solanki, V. K., Katiyar, S., Semwal, V. B., Dewan, P., Venkatesan, M., and Dey, N. (2015). Advance automated module for smart and secure city, in *First International Conference on Information Security and Privacy (ICISP 2015)*, G.H. Raisoni College of Engineering & Information Technology, Nagpur, India, December 11–12, Procedia Computer Science, Elsevier.

Solanki, V. K., Venkatesan, M., and Katiyar, S. (2017). Conceptual model for smart cities for irrigation and highway lamps using IoT, *International Journal of Interactive Multimedia and Artificial Intelligence*.

Solanki, V. K., Venkatesan, M., and Katiyar, S. Think home: A smart home as digital ecosystem in circuits and systems, *Circuits and Systems*, 10(7).

STAMFORD, Conn. (September 16, 2015). https://www.gartner.com/en/newsroom/press-releases/2015-09-16-gartner-survey-shows-more-than-75-percent-of-companies-are-investing-or-planning-to-invest-in-big-data-in-the-next-two-years. Accessed October 12, 2015.

Stephens-Davidowitz, S., and Pinker, S. (2017). *Everybody Lies: Big Data, New Data, and What the Internet Can Tell Us about Who We Really Are*. New York: HarperCollins.

Stucke, M. E., and Grunes, A. P. (2016). *Big Data and Competition Policy* (p. 15). Oxford, UK: Oxford University Press.

Trifu, M. R., and Ivan, M. L. (2014). Big data: Present and future. *Database Systems Journal*, 5(1), 32–41.

Turkington, G. (2013). *Hadoop Beginner's Guide*. Birmingham, UK: Packt Publishing.

Vohra, D. (2016). *Practical Hadoop Ecosystem: A Definitive Guide to Hadoop-Related Frameworks and Tools*. Berkeley, CA: Apress.

Wang, L., and Jones, R. (2017). Big data analytics for disparate data. *American Journal of Intelligent Systems*, 7(2), 39–46.

Wang, L., and Jones, R. (2017). Big data analytics for network intrusion detection: A survey. *International Journal of Networks and Communications*, 7(1), 24–31.

White, T. (2010). The Hadoop distributed filesystem. *Hadoop: The Definitive Guide*, 41–73.

Zikopoulos, P., and Eaton, C. (2011). *Understanding Big Data: Analytics for Enterprise Class Hadoop and Streaming Data*. New York: McGraw-Hill Osborne Media.

10

Impact of Big Data in Social Networking through Various Domains

Abhishek Kumar, Srinivas Kumar Palvadi,
T. V. M. Sairam, and Pramod Singh Rathore

CONTENTS

10.1 Introduction

Data is being collected from different regions of the Internet, and one main area we can focus on to collect various types of content-oriented data is from social networking. Here we can have a huge impact on various data domains because of the different sorts of data being collected from social networking. We can consider data gathered from social networking as posing a big data issue: How can we best use and implement this data? The IoT (Internet of Things) is another source of information generation and retrieval from various sensors in various domains, and implementations using MapReduce and other big data tools like Hive, Pig, etc. are being considered, but few particular uses or applications have been identified that specifically focus on social networking. This chapter will mainly focus on what implementations are being done and the impact of big data analytics on social media in various domains.

The day-by-day implementations in social networking sites, updates given to social networking software, and lots of real-world modifications in daily life will all create big data. Yes, it's true. It's not a generalized information statement. To further consider this statement, we can look back to the initial stage of social networking, how it has developed, and how it is changing our lifestyles. In this scenario we can consider social networking with regard to its creating large amounts of unstructured and semi-structured data. This is very tough to process. We have the MapReduce model for trying to identify the requirement of maintaining a piece of data in a repository, and the scenario here is to discuss how big data will impact social networking on various domains.

This chapter delineates our mining and examining information from an electronic frameworks perspective. Our strategies for information managing contain three phases and are informed by our contemplation of two extraordinary phenomena: colossal information and scattered figuring. In any case, the second stage is information mining from web frameworks organization and its filtration or gathering, which results in a few respectable datasets that have information essential to understanding our task. Information mining is performed by a crawler which is organized in the big data framework, like MapReduce display for scattering calculations and which we finished utilizing Hadoop structure. In the last part of this chapter little datasets are examined using push models. The entire information examination process is formalized in the composite apps, which are kept in running condition by the scattered-figuring-based cloud organization CLAIRE.

Another part of this chapter delineates examination of individuals who elucidate claims in electronic frameworks' organization for the Internet of Things. Because here we need to apply a framework to implement the Internet of Things. We demonstrate our thoughts of utilizing electronic frameworks organization as an extra information hotspot for inspection and displaying of unlawful exercises in the general public eye. Types of progress in mining and analysis are related to delineating clients who illustrate their drug use. The main attributes are not hidden and there is transparency with quality also. This chapter likewise depicts guess appear for the position of pharmaceuticals to use by individuals which consider different parts, as the broad-scale condition of the majority and individual attributes of occupants. Consequences of online social frameworks' organization examination, for example, a position basic to the pharmaceutical topic or characteristics of clients who utilize meds, can be used to expand exactness of this model [1].

Regardless of the generally late appearance of the term "big data" in 2008 [4], this space of science pulled in the gigantic idea of business and the adroit world. Various undertakings require examination of great measures of information, for instance, testing the Large Hadron Collider, air expansions, and creation recommendation structures for web associations; however, the need for interpretive blueprints is particularly fundamental for the business. The fiscally sharp reaction for the company is to utilize haze structures, which will give coordinate assets and undertake associations. Shockingly, information association and its examination in the hazes give an amazing measure of inconveniences, and at present, there is no absolute best approach to manage interfaces with them. What the web gets sorted out is ceaselessly being utilized as a bit of the scholarly gathering and business to break down the specific strategy of this present reality; for instance, securities exchange wants, viral publicizing, the occasion of flu, and seeing potential new risky medication reactions. They are in the same way a lot of investigations consider drug enslavement, for instance, about the communication between needy-sharing cases and honest to goodness easygoing affiliations and social affiliations influence on the progress to blending drug utilize which considers individuals physically subject to drug and their association sort out. This concept makes the decision that tries to use information from online social frameworks' organization, i.e., the virtual machine, to think about sedative utilization. The perfect fundamental position of this work is that it gives bits of data on the scarcely noticeable social event of individuals who now and again utilize sedatives or aren't dependent on them.

The later part of this chapter deals with the background operations to date in this concept of managing data in various domains; later we discuss experimental analysis; next, data analysis and data management; next, we will discuss future work and conclude the topic with acknowledging the most important people and mentioning the references considered.

10.2 Background

Research of pharmaceutical use touches various demands; a few of them are tended to in this concept: estimation and lack of a solution usage position by the majority, drawing up a run-of-the-mill position for a man who utilizes drugs. In any case, different unmistakable solicitations which are appropriately a pure human science can be asked for: What are the causes behind a man utilizing medications, and how to decrease a position of unlawful use. To get an opportunity for the solution of these demands, different parts of pharmaceutical usage ought to be isolated. Also, as with other social techniques, enduring use examination gives an excellent measure of challenges in light of the way that many scarcely formalized components ought to be considered. For instance, remarkable sorts of solutions have grouped positions of utilization among various ages' and sexes' social events. Correspondingly, social, subsocial, and ethnographic parts of the majority and the monetary situation of the district have an impact on medication use.

Moreover, the position of the pharmaceutical use changes over time in light of the developments in various segments: social rebuffs, which effect the utilization of any medicine; law endorsement workouts; transparency of headway; and differing changes in a specific culture [2]. It is essential to equalize the domains into categories in light of the pharmaceutical utilization position. This equitation changes, as the time since the drug user can change his or her view of drug use and exchange in between get-togethers.

Generally we see the categories of parties of individuals which are hereafter indicated by Roman numerals: (i) not identified with steady usage and not powerless in their drug use; (ii) needing to cure misuse of drugs, as they start to restrain misuse, perhaps by associating with others of certain interests, ages, social parties, and so forth; (iii) sedative clients who don't have drug dependencies; (iv) clients having a mental dependence on drugs; (v) buyers having a physical reliance on drugs. It ought to be seen that it is possible for a person to have both mental and physical dependencies on drugs, and in this situation we would suggest he join the fifth (v) gathering. Showing the arrangement utilizes expectations to gauge of the sizes of the depicted get-togethers or how individuals exchange information, starting with one of these social affairs, then onto the accompanying areas of interest.

Following the depiction of remedies, the laws on controlled psychotropic substances categorize all meds into five classes: fortifying drugs, psychedelic meds, stimulants, sedatives, and sedative rest-inciting drugs. Courses of the danger of physical dependence come later: solutions of considerable peril include sedatives, two or three stimulating meds (heroin, LSD), or medications of chance: stimulants, depressants, low-risk drugs. The utilization of meds causing physical reliance impacts people to alter their advancement as for their need to get standard estimations of meds and expend cash on a pharmaceutical. So also, moderate use can change an individual's way of life, as his new associates will similarly share his interests. Each prescription made has a target amassing out of individuals with a particular way of life. Sedatives are prominent among individuals who have physical as well as mental reliance. Also, fortifying pharmaceuticals are standard among individuals who are amped up for muscular development and for changing personality. The subject of finding these outside parties for various drugs is incredibly intriguing. Explaining this demand can help in creating a solution that will decrease illicit medication use.

It is additionally critical to consider distinctive medication compositions independently, because they impact how an individual acquires a mental or physical dependency differently, depending on the medication used. Following the order of drugs, described by the Controlled Substances Act, all medicines falls into five classes: hallucinogenic medicines or psychedelic drugs, stimulants, sedatives, depressants, and entrancing narcotic medications. Classifications of the peril of physical dependence are next: medications of high threat, such as sedatives, a few psychedelic drugs (heroin, LSD); moderate-hazard, such as stimulants and depressants; and low-hazard drugs. The utilization of medications causing physical reliance influences people to modify their movements in connection with their need to get standard measurements and appearances of medications and necessarily to spend cash on a drug acquisition. Also, drug utilization can change an individual's way of life, so his or her new social connections will likewise share his interests.

Each drug written about in social media is targeted at a gathering of individuals with a particular way of life. Sedatives are famous among individuals who have physical and mental dependencies and who have compulsions to medicate themselves. And use of psychedelic drugs is prominent among individuals who are keen on extending and changing the human mind. The topic of how one finds these goal-oriented gatherings for various medications is very interesting and still uncertain. Researching this question can help to eventually alleviate the impact of social media medication promotions, which could ultimately reduce medication abuse. Russian investigations are very tentative as a result of the unlawful and mysterious nature of illicit medication use. Only a few wellsprings of data exist, which, shockingly, are one-sided and unfinished. Direct interviews about drug-use encounters are conceivable, yet mostly untested, and the information available is primarily regarding "soft drug" use. It's likewise conceivable to track the number of individuals who have acquired a physical reliance on the medications and who have been "enlisted" by narcologists. This information source can give information only about gatherings categorized as (iv) and (v) earlier in this section, and the information available is strictly regarding individuals who utilize opium. Low-risk drugs are not at all discussed in this information. The most profitable wellspring of data concerns legitimate insights about the seizure of unlawful medicines, since it indicates the condition of the medication showcase, which, like this study, correlates with request and utilization of the medications among the populace.

We advise using web-based social systems administration as a further wellspring of data about meds: to begin with, of everything they can give information about on the position of interests to the meds' subjects, and furthermore, they would provide data that have the capacity to give information about different subsocial and mental characteristics of people who use drugs. This information can be used as additional components to demonstrate the position of the drug use as well as to make the data much more productive.

Our work relates to the gathering of work on online abnormality as it has been investigated by the product building and by logical showing approaches. A few late works focus on perceiving social spammers. Social spammers, as per this gathering of the task, are customers composed either with the help of individuals, who uses social systems' administration goals as well as particularly the social relationship to propels things, advance events, or post vain and also graceless comments. Lee et al. thought about social attackers in online social frameworks. In considering this, they sent social nectar pots for harvesting overwhelming spam profiles from individual-to-individual correspondence gatherings and made spam classifiers using automation techniques in perspective of a grouping of features.

So likewise, Domingos [18] advanced a mechanism for distinguishing comment spam by using the usefulness position of a comment; it showed that spammer's remarks have low information positions. Hu et al. in like manner proposed a structure for social spam area, in perspective of social system content, achieving a high exactness of revelation. Our emphasis isn't on social spammers, be that as it may; instead it's on practices related to vandalism and awful direct content which could be delivered by social spammers. To oversee demolition and bots, a couple of devices exist. Automated bots, channels (i.e., manhandle channel), as well as changing accomplices are all mean to discover vandalism acts.

Such gadgets work by methods for predictable articulations and physically composed oversee sets. Regardless of the way that clients post, these systems can't elucidate or predict which harming and monstrosity practices will happen and are routinely obliged to rely on lingo recognition. Finally, our concept makes the gathering of tasks on free-riding in dispersed structures. Distributed networks are proposed to access customers to interface with others as well as offer information. Practically identical to online gatherings, customers are permitted to get the chance to contribute as much as needed, and few controls are set up. Therefore, in p2p (person-to-person) structures, companions may misuse their relationship by mishandling other associates' advantages, declining to share guaranteed assets, sharing broken or debased resources, etc., depleting the association without contributing to it.

In online groups, the quality of the gathering is intensely based on particular allies' reactions to narrow-minded conduct, which they may emulate or withdraw from. Teaching parts can likewise be set up, regardless of the way that these are every now and again considered not to be extremely fruitful. To deal with these issues, the most customary plan is the execution of impetus-based instruments [3]. Driving forces are connected to guarantee online social affairs, whereby end customers are given unique parts and advantages as a reward for good behavior. From a theoretical perspective, our work can be set concerning the Detroit show in which singular person changed her determination progressively and to a constrained degree through pantomime. In that model, a discrete-time Markov chain shapes the underlying conduct.

By a preoccupation point of view, Morris thinks about behavior infection by coordination diversions. Our task can be considered as the prisoner's dilemma; be that as it may, our guideline result is appropriate for a more broad class of preoccupations. Likewise, our work is persuaded by work crafted by Jackson et al. when we consider a proposed appearance on the closeness of framework changes.

In this work related to big data analytics and IoT, we propose a model for the spread of variation in online casual groups. We don't propose parts for teaching and reward of worsening direct; rather we demonstrate the innate get or incident for a customer in the framework who shows held, crack lead, and in a like way, how this direct and also observed get or mishap coming to fruition on account of this direct impacted the approaches of various customers in the structure after some time.

Relational associations have pulled in mind-boggling excitement in the last few years, as it were, in light of their plausible essentialness to various social systems; for instance, information taking care of (1), scattered interest and spread of social effect. For quite a while, regardless, social specialists have also been enthusiastic about casual associations as intense techniques in themselves [4]: the individuals making and deactivation of social ties, in this way adjusting the structure of frameworks in which they partake. Social mastermind improvement is a psyche-boggling process in which various individuals at the same time attempt to satisfy their targets under various, conceivably conflicting, limitations. For delineation, individuals consistently coordinate with others like themselves—their inclination defining them as homophiles—and attempt to refrain from conflicting associations while abusing cross-cutting circles of colleagues. Nevertheless, the affirmation of these points is at risk to the spatial and social proximity of available others.

In some conditions in which the individuals are benefited by pleasing associations, they may emphasize introduced ties: those having a place with locally thick gatherings [5]. For example, they may pick new partners who are buddies of allies—a methodology known as a triadic conclusion (12). They might in any case in like manner search for how to view the novel information, resources, and hereafter have the advantage of having access to ranges—relationships outside their drift of associates—or by spreading over assistant holes accurately between other individuals who don't have any colleagues in common with each other. Finally, social ties may deteriorate for certain causes, for instance, when they are not maintained by various relations, or else battle with them.

Diverse social and various positioned settings are an observational issue, requiring longitudinally (i.e., assembled after some period) organized information joined by data about individual characteristics and get-together affiliations. However, longitudinal framework data is extraordinary, and the best-known representations are for little social events. Late examinations of substantially greater frameworks, in separate studies, have tended to focus on cross-sectional (i.e., static) examination or they have underscored the correspondences between people or their get-together affiliations, yet not both.

The examined framework has a dataset that was made by mixing three specific, somehow related data structures. In any case, we organized a registry of email relationships in masses of 43,554 students, faculty, and staff members of a generous school using a traverse of one year.

For every email message, timestamp, and sender, the number of persons were recorded. Secondly, for comparative masses, we collected information showing an extent of individual properties. Later, we procured complete plans of the classes went to and informed, independently, by understudies and educators in each semester. For security protection for all individuals as well as social affairs, identifiers were mixed [6]; we can choose, for example, paying little mind to whether two individuals are in a comparative class.

Since in a school setting, classes' support gives essential opportunities for eye-to-eye participation, we used classes to address the change association.

Our usage of email correspondence to a fruitless principal arrangement of social ties is maintained by late examinations declaring the use of mail by close-by gatherings of companions is unequivocally compared with very close telephone associations (23,24). Individuals and social events of individuals may differentiate in their email utilization; consequently, enlistments drawn on a little case of granting sets may be baffled by the whimsies of particular characters besides associations. In any case, by averaging more than a colossal number of such associations, we analyzed that our results will address only the broadest regularities overseeing the beginning and development of social correspondence. To make sure that our data does, as a general rule, reflect social trades limited to uncommonly selected mailing records and several mass mailings, we filtered through messages with more than four recipients. Ensuring for isolation, there were 15,584,421 messages exchanged by the customers in the midst of 355 days of discernment,

Continuous social connections deliver spikes of an email trade that can be watched and, what's more, tallied [7]. The more grounded the relationship between two people, the more spikes will be watched for this specific match, by and large, inside a given time interim.

We estimated an immediate quality of a connection between two people and j by the geometric rate of two-sided email trade inside a window of t 0 60 days. The prompt system anytime incorporates all sets of people that sent at least one message toward every path during the previous 60 days. Utilizing everyday arrange approximations, we figured the following:

1. Most limited way lengths DID
2. The number of shared affiliations SIJ for all sets of people in the system on 210 back-to-back days spreading over the vast majority of the fall and spring semesters

   By recognizing new ties that show up in the order after some time, we can process two arrangements of measures: (1) cyclic conclusion (2) central conclusion inclinations.

For some predefined estimation of DID, cyclic conclusion predisposition is characterized as the observational likelihood that two beforehand-detached people who are separate DIDs separated in the system will start over again to connect or tie. In this manner, cyclic conclusion sums up typically the idea of triadic conclusion, i.e., the arrangement of cycles of length three. Similarly, we characterize central conclusion inclination as the observational likelihood that two outsiders who share a collaboration center (in the present case, a class) will shape another tie. Since class participation is essential for the most part for understudies, the outcomes on nearby and cyclic conclusions are introduced here for a subset of 22,611 graduate and college understudies (25).

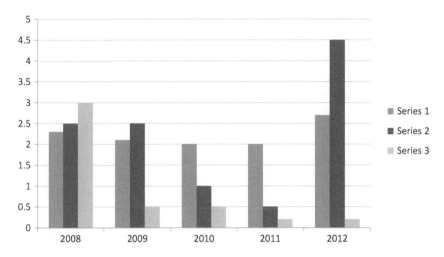

FIGURE 10.1
Presence of drug as a presence of representation.

Figure 10.1 demonstrates without a mutual concentration (i.e., class), cyclic conclusion decreases quickly in quality inside, inferring that people who are far separated in the system have no chance to associate and subsequently are probably not going to forge ties. For instance, people who are isolated by two middle people (DIJ 0 3) are around 30 times more averse to starting new ties than people who are separated by, as it were, one middle person (DID 0 2). Figure 10.1a circles, in any case [8], exhibit that when two people share no less than one class, they are by and large 3 times more inclined to associate on the off chance that they likewise share a colleague (DID 0 2), and around 140 times more probable on the off chance that they do not (DID 92). What's more, Figure 10.1b demonstrates the exact likelihood of tie development increments within quantity for ordinary colleagues both for sets with and without shared classes, getting to be noticeably free of shared affiliations for the vast numbers of regular colleagues. Figure 10.1c shows comparable data for shared classes, demonstrating that while the impact of a separate shared class is generally tradable with a solitary ordinary colleague, the nearness of extra associates has a more noteworthy impact than extra foci in our informational index. These discoveries suggest that even an insignificantly exact, generative system model would need to account independently for (1) triadic conclusion, (2) central finding, and (3) the intensifying impact of the two inclinations together.

Our information can likewise reveal insight into possible thoughts of tie quality (13) and trait-based homophiles. We discovered the probability of triadic conclusion expands the normal tie quality between two outsiders and their shared associates are high, which underpins the ordinarily acknowledged hypothesis. By differentiating homophiles, singular ascriptions appear to play a weaker part than may be normal. Of the properties we considered in this and other models—status, sex, age, and time in the group—none significantly affects the triadic conclusion. The critical indicators are tie quality, various common associates, shared classes, the communication of shared colleagues, and status check, which we characterize as the impact on the triadic conclusion of an intervening person who has an unexpected status in comparison to both of the potential colleagues. For instance, two understudies associated with a teacher are more averse to frame an immediate tie than two understudies related to another understudy, ceteris paribus. We think, be that as it may, that status hindrance might be a marker of an imperceptibly nearby conclusion of past class attendance. Thus, albeit homophile has frequently been seen as a singular characteristic in cross-sectional information [9], these impacts might be for the most part backhanded, working through the basic requirement of shared foci (10; for example, choice of courses or extracurricular exercises.

Our outcomes additionally have suggestions for utility in cross-sectional system investigation, which depends on the presumption that system features for premium in equal manner (4) demonstrate the distinctive system measure to display fluctuating positions of strength after some time, and for the smoothing window T. Ordinary vertex degree *bkà*, partial size of the biggest segment S, and mean briefest way length L all show occasional changes and create distinctive estimations for various decisions of t, where *bkà* is particularly delicate to t. The bunching coefficient C (28), in any case, remains practically steady as *bkà* changes, proposing, maybe shockingly, that the midpoint of nearby system properties are steadier than worldwide properties, for example, L or S. In any case, these outcomes recommend that as long as the smoothing window is picked suitably, and we take mind to abstain from gathering information in the region of exogenous changes (e.g., end of the semester), normal system measures stay stable after some time and in this way can be recuperated with sensible loyalty from organizing depictions.

The relative steadiness of normal system properties, in any case, does not infer proportionate dependability of individual system properties, for which the observational picture is more complicated. On the one hand, we find that dissemination of individual-position properties is steady, with similar admonitions that apply to midpoints. For instance, the state of the degree appropriation $p(k)$ is moderately steady over the span of our dataset except amid simple spells of diminished action, for example, winter break. Practically equivalent to results apply to the idea of Beak ties [10]. The dissemination of tie quality in the system is steady after some time, and scaffolds are, all things considered, weaker than installed ties are consistent with. Be that as it may, they don't hold their crossing-over capacity, or even stay feeble, inconclusively.

Our outcomes propose that conclusions relating contrasts in result measures, for example, status or execution to disparities in a unique system position, ought to be treated with an alert. Extensions, for instance, may, in reality, encourage dissemination of data crosswise over whole groups. In any case, their precarious nature suggests that they are not owned by specific individuals indefinitely; in this manner, whatever favorable circumstances they give are additionally brief. Besides, it is unclear as to what degree people are prepared to deliberately control their positions in a vast system, regardless of whether that is their aim. Or maybe, it gives the idea that individual-position choices watch out for leveraging. Sharing central exercises and associates, for instance, enormously improves the probability of people getting to be noticeably associated, mainly when these conditions apply all the while [11]. It might be the situation that the people in our populace—for the most part understudies and personnel—don't deliberately control their systems since they don't have to, not because it is unimaginable. Along these lines, our decisions concerning the connection among nearby and worldwide system elements might be particular to the specific condition that we have examined. Similar investigations of corporate or military systems could help enlighten which highlights of system development are blind and which are particular to the social, hierarchical, and institutional setting being referred to. We take note that the strategies we presented here are non-exclusive and might be connected effectively to an assortment of different settings.

We close to stress that the understanding of tie arrangement procedures at informal organizations needs information on social associations as well as shared affiliations. By the fitting informational collections, vague guesses can be tried straightforwardly, and conclusions beforehand given cross-functioning information can be approved.

The utilization of huge information, for example, enormous information examination on the Internet of Things, is more critical. The person-to-person communication benefit, as one of these applications, has gotten full consideration. Interpersonal interaction administrations are not constrained to the individual-to-individual correspondence but rather reach out to a more extensive territory including individual-to-thing correspondence and thing-to-thing correspondence. In this manner, it is likewise called enormous long-range interpersonal communication administration. In the interpersonal organizations, participation assumes a crucial part of the advancement procedure of species from cell living beings to vertebrates. Be that as it may, understanding the development of collaboration in the advancement hypothesis stays a test to date.

Additionally, the developmental diversion hypothesis has been considered as an essential research structure to describe and comprehend the collaboration component in the frameworks comprising of people. A great deal of consideration is being paid to the investigation of the developmental progression of pair-wise communications, for example, the prisoner's dilemma game, the snowdrift amusement, and the stag-chase amusement. In these models, being a deserter is continuously superior to being a co-operator. In any case, the two players like co-operators are always superior to the two players as deserters. These diversion structures mirror the circumstance of enthusiasm for reality. The amusement display, the system topology, and the transformative manage are viewed as three key elements of a developmental diversion in a system. As of late, these amusement models in typical systems, for example, little world systems and sans scale systems, have been seriously considered. Under a few conditions, social predicaments include more significant gatherings of interactional people as opposed to sets [12]. Open products amusement was acquainted with clarifying these questions.

Even though childishness and intensity are inalienable characteristics of human instinct, people will coordinate if legitimate conditions are accessible. Participation disappointment brings about the abuse of open products, for example, natural assets or social advantages, by turncoats, who receive rewards to the detriment of co-operators. The "deplorability of the lodge" compactly portrays such a circumstance. A yearning-actuated reconnection instrument has been presented into the spatial open-merchandise amusement [10]. A player will reconnect with an arbitrarily picked player if the result obtained from the gathering fixated on the neighbor doesn't surpass the desired position.

The long range of SNS, as the result of Web 2.0, is another mechanism, in that "cooperation" is viewed as a center. Entirely, SNS ordinarily alludes to interpersonal interaction sites, including its inferred items. Those clients, who are data beneficiaries as well as makers, dischargers, and disseminators of data, turn into a basic piece of the procedure of arrangement and advancement of general supposition in systems. These days, informal online communities have pulled in a large number of clients. In informal organizations, clients connect with others, manufacture connections, distribute posts or answers, and examine points. Along these lines, the development of informal organizations is advanced by clients' activities.

As a sort of open asset in interpersonal organizations, data is just created by a few clients and shared among clients in a bigger degree. Turncoats get the advantage with no cost and in the long run, drive the system into a dull climate. Chairmen and administrators of interpersonal organizations fell into a predicament: how to elevate clients to impart data and convey to each other much of the time. To take care of this issue, in this chapter, we propose a prowler diversion display [13]. The prowler diversion is a sort of the transformative open merchandise amusement. Notwithstanding the highlights of general society products amusement, this model additionally presented the factor of original impetus relying upon his degree. We observe that the individual methodology to be picked isn't applicable to the client's degree yet to a motivating force steady of the whole system. The re-enactment comes about demonstrated that individual procedures asymptotically took after three distinct practices as indicated by the dynamic association of the people. Dynamic clients rose amid the developmental process with an impetus. Without a motivation, dynamic focal clients could scarcely influence the conditions of their neighbors and may even move toward becoming prowlers because of the immense number of sneaking neighbors. Substantial commotion diminishes the impact of the high motivator and confuses systems. On the off chance that the disorder is ceaseless, dynamic clients will continuously lose interest and leave the system.

10.3 Experimental Analysis

Weibo, a Chinese smaller-scale blogging organization, has ended up being a champion among the most regular online long-range informal communication on the Internet, and its enlisted customers and month-to-month dynamic customers in 2015 are respectively 503 million and 212 million. Its effect develops from one-of-a-kind spaces, for instance, news and redirection to fresh territories, for example, amusements and travel. Weibo is reliably the wellspring of predominant focuses with general conclusions toward them.

Here, a significant measure of information was evacuated within our Internet bug included data regarding customers and their information. Following 30 minutes of social occasion from Weibo, 422,171 customer profiles and around 2.4 million presents from May 2014 to July 2014 were downloaded. The typical degree was 26.3, and the gathering coefficient was 0.14. The degree dissemination agreed to a power law, a general property without scale frameworks. By then we viewed the development status of each customer every other hour on this site.

In light of the casual associations' options of large packing and much accessibility, charming issues can distribute to immediately complete a wide range. Customers in the central position are ordinarily prepared to share the new resource in light of the way that their got effect is essentially more noticeable than their extraordinary effect got from them in advance of given resources. The effect of a customer contains three dimensions [14]. The very beginning performance is the active position, which is evaluated by the measure of the edifying substance made by a customer; the second one is the transmissibility, the ordinary sent conditions per post of a customer; the final one is the extension, which is for the most part assessed by the number of customers tolerating his post, including his quick fans and indirect customers getting his post sent by various customers. Figure 10.1 exhibits the assortments in the three edges with customers' degrees.

A customer with a large degree, 107, will appropriate a couple of posts every day since he or she may be a major name, especially a pop star or a film star. Each of his or her exercises mixes the overall public interest. In perspective of notoriety [15], the amount of sent conditions per post and the amount of customers getting his post is far more significant than various customers in the framework. People focused on these stars are far more inclined to share the status among their associates. A client with a small degree, 106 or 105, perhaps a VIP in particular fields, more often than not distributes one post each day to keep up his position in the area. A dynamic client in a little group, with a view of 500 to 104, is more inspired than the previous to distribute presents since he needs to advance his or her notoriety. A dynamic client regularly distributes more than one position each day to be in contact with this companion; the rest all of them are often. An idle client with a value under 101 doesn't provide the rationale to distribute any post. He generally sees other individuals' posts and once in a while advances one of them [16]. In this paper, another term, prowlers, is given to this kind of client. A post from a prowler can't excite consideration and is once in a while sent by different clients. Likewise, some clients getting a client's post is about, to an extent, his degree. Considering all the three viewpoints in general, a client's impact is predictable with his degree. A high degree dependably brings about a critical impact.

We claimed two clients from the tremendous number of amassed clients. The two clients utilize the administration pretty often and don't prowl or leave the administration in June 2015. One of them has an expansive degree (1, 7104, 701); the alternate has a moderately small degree (533). As appeared in Figure 10.2, the two clients are enormously affected by the movement statuses of their neighbors. The client in Figure 10.2a, who has an expansive number of neighbors, is a focal client. In the meantime, the client in Figure 10.2b is a standard client. The majority of the neighbors of the focal client may simply see the post from big names and don't distribute any post without anyone else.

At the point when dynamic neighbors turn out to be less, whatever is left of them are focal clients, who much of the time distribute presents to keep up their fame in the groups. The green squares in the upper part of Figure 10.2a and b demonstrate the client's movement statuses across the month. The focal client is constantly dynamic with no intrusion due to his focal position in the entire system. His fans will know his everyday life and working circumstances [17]. High benefit for each of his presents spurs him on to stay dynamic. Conversely, the typical client changes his movement status as indicated by his neighbors' past statuses.

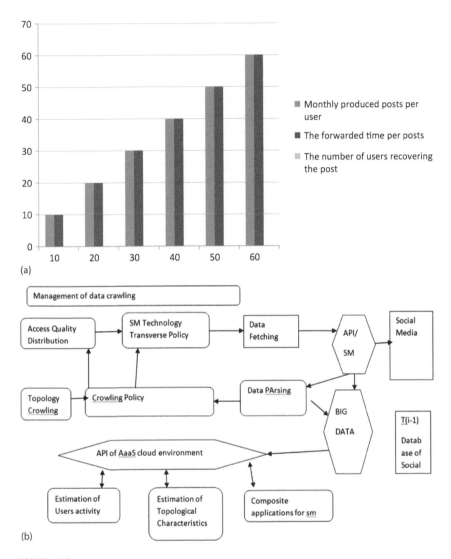

FIGURE 10.2
(a) Users influence versus users degree. (b) Iterative crawling procedure.

10.4 Data Management

This area portrays our administration of information from online networking, which brought together the approach for explaining logical undertakings. Since online networking contains a gigantic measure of information, we utilized big data paradigms to mine and break down. Right off the bat, information from online networking is mined utilizing our crawler, which stores it into the Hadoop group. Furthermore, the enormous volume of mined information is sifted and totaled all together to get generally little datasets of data that is pertinent to the illuminating assignment [18].

At last totaled data is utilized as a contribution to composite applications, which perform last and refined information examination. To organize the computational procedure of the composite application, we utilized the AaaS model, which is actualized in our condition for the appropriated processing-based cloud platform CLAIRE. The application works by information which is acquired from a big data group using created API, and counts are performed by the computational module which can be flawlessly incorporated by CLAIRE. Subsequently, it provides the approach to digging as well as investigation information from online networking.

Information from online networking is acquired from the API-supplied social media destinations. Web-based social networking contains diverse sorts of information: client data, connections between clients, produced by clients' substance, and so forth. Every datum is composes, more often than not, get to by discrete API [19] and every opus forces limitations to the measure of gets to. This prompts the difficulty to gather in reasonable time all information put away in web-based social networking.

However, even with these confinements, measures of information are big. Likewise, information writes can be organized (e.g., client profiles, interfaces among clients) and unstructured (e.g., user's interests, posts or pictures). To effectively work with the large and unstructured information we utilize the Hadoop framework, which executes MapReduce to show disseminated calculations and other associations with its advances. Another imperative normal for the info from online networking is its "sparseness," which lies in the way that an exclusive little measure of information is essential to taking care of an issue. This means that huge unique details mined from online networking ought to be sifted, mapped, or amassed to some small dataset that is utilized as a part of a further investigation.

Hadoop structure comprises from two primary parts: HDFS and NoSQL. The circulated record framework HDFS component performs information preparing in MapReduce show. HDFS utilizes NoSQL information display in which records contain key and esteem parts of self-assertive write. Records shapes tables, which on the physical position are put away by parts (lumps) on a few machines that work under the Hadoop system and processed setup. HDFS is exceedingly adaptable, reliable, and in the meantime adaptable. One of its benefits is its capacity to store information on the physical position in various organizations. We utilized Google Protobuf binary serialization system to save data recovered from interpersonal organizations' information. Protobuf additionally streamlines handling situations with information design evolving. To build execution of the HDFS [20] we pack information using Snappy pressure arrange, which is intended to quick pack and decompress data; however, it has an average compress proportion.

In the issues we confronted, diverse information composes were amassed independently yet further aggregated small datasets were viewed as together, keeping in mind the end goal to get the last outcome. Thus the last analysis required the work of a few stages, each speaking to some fundamental calculation or investigation. This enlivened us to formalize this examination as work processes or composite applications that keep running on CLAIRE. CLAIRE [21] is a distributed processing-based cloud stage of the second-stage that actualizes worldview AaaS (Application as a Service) and that gave consistent reconciliation of independent programming modules.

Our approach is shown with two illustrations. Initial one depicts an iterative calculation for mining information from online networking. Every emphasis by solicitations to online networking is prepared and reactions are composed to HDFS. This information can be explicitly investigated on the Hadoop group; for instance, to assess the movement of clients or to get topological qualities of the systems.

This application keeps running in disseminated computational condition CLAIRE. Consequences of this investigation are utilized to control creeping process and to make arrangements for solicitations that will be handled with the following emphasis. Crawler control system is likewise actualized as the application that understands the subtasks: filtering of items, shared conveyance between as-of-now slithered protest of web-based social networking, and picking which new questions ought to creep on the following emphasis.

Another illustration (see Figure 10.3) depicts the application that is utilized to discover groups of clients that are keen on the same themes [22]. In the initial step, information from the interpersonal organization is mined, and afterward, information related to clients' interests and labels that they used to check posts are removed. These means are performed

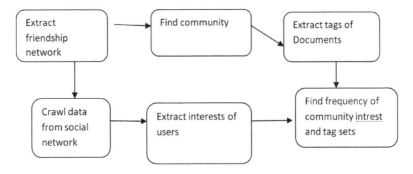

FIGURE 10.3
Examples of workflows that run on distributed computing–based cloud platform.

utilizing Hadoop bunch. In the meantime, fellowship arrange is broke down and the group structure is extricated. Lastly, for every group, the most continuous interest and label sets are disengaged. These means are acknowledged as independent and can be reused for some unique systems. Union of these means in the work process shapes composite applications, which can be further straightforwardly utilized by customers of the framework.

This segment depicts an examination of clients that have online journals at live journal SNS that can be the used in a model of medication use in the public arena. We examined clients who expound on drugs on their sites and made their attributes by dissecting interests. Properties consist of the arrangement of points of interest that emerge this gathering of clients radically high or low frequencies. This permits making the mental picture, which portrays highlights that distinguish individuals from this gathering from every single other client and to decide their way of life. Descriptions in this area approach can be additionally used to make attributes of any gathering of clients, and they have functional significance in the event of breaking down the gathering of clients [23] with the normal component, which is fairly hard to acquire in reality.

It ought to be noticed that arrangements of individuals who utilize sedatives in reality and those who expound on them in web-based social networking are not the same; in any case, they should cover. The client may have many motivations to expound on drugs: curing sedative dependence might be his or her calling; he or she may attempt to attract thoughtfulness regarding the issue of the expanded position of medication use; a client may expound on not utilizing drugs and a solid way of life; a client may essentially utilize sedative or tranquilizer slang as a joke. In any case, we trust that in the colossal measure of the information from web-based social networking is data about clients who offer their experience of medication utilization. What's more, by breaking down the client who expounds on drugs, it is conceivable to extricate data about individuals who utilize drugs.

To distinguish clients who expound on drugs, we utilized the lexicon of the authority and slang catchphrases and key phrases. Every catchphrase was relegated to a few gatherings relying upon semantic significance. We distinguished after gatherings of words: weed, crude, infusions, arrangement. Key phrases are characterized in light of the predefined rules, which thus depended on the semantic gatherings of watchwords. For instance, we utilized after principles to naturally make the arrangement of key phrases: "drugs" + "methods for utilizing"; "drugs" + "readiness"; "drugs" + "impacts" and "medication" + "tranquilize equivalent word." We likewise did the manual amendment of the key phrases with a specific end goal to add new expressions or to expel expressions important to sedative subjects. We additionally physically split watchwords into three gatherings, relying upon their "quality" of having a place with the medication subject [24]. Each gathering has a similar weight, which is utilized to gauge the importance of the content to the medication subject. We influenced a key phrase to a weight bigger than the total of its watchwords' weight to demonstrate that key phrase is the more grounded flag of record pertinence to the medication subject. The heaviness of the record is ascertained as an aggregate of all catchphrases and key phrases [25] that are found in it. The top 20% of records with the biggest weights are thought to be pertinent to the medication subject. On the off chance that the client has no less than one archive pertinent to the medication subject, then he is thought to be keen on the sedative or tranquilizer theme. Our word reference comprises 368 catchphrases and 8359 key phrases. We dug information from around 100,000 haphazardly chose clients from a live journal social network site. For every client, we know his or her last 25 posts, his or her incoming and out-running associations with different clients, information that he or she indicated in his or her profile and interests—catchphrases that portray client's regions of interests (e.g., music, movies, PCs or game). Utilizing the strategy portrayed above, we recognized 16553 clients who expound on drugs. We will additionally call this gathering of clients the medication group.

Give I={I1, I2,...,IM} a chance to be a set of interests determined by all clients in the informational collection. Each interest is a depicted arrangement of clients that characterized it in their profiles, and for every client, it is known whether he or she is from the sedative group or not. At that point, for each interest it ought to be resolved whether it shows up fundamentally pretty much frequently in the medicating group. Give I={I1, I2,...,IM} a chance to be this arrangement of noteworthy interests, in light of which we will draw up normal for clients from the sedative group.

To verify if an interest is noteworthy, each interest is portrayed by four esteems, or "likes," that the aggregate positions of clients in the medicating group indicated in their profiles. The number of clients from the sedative group who don't determine in their profiles and a comparable two likes for clients who are not from the sedative group are added. In light of these four likes a 2 × 2 possibility table is made, for which the Fisher Exact Test is connected. The Fisher Exact Test checks the theory that interest is critical and returns for each interest a relating value. To supply a manual investigation of the set we consequently gathered premiums' more broad subjects. Statistical examination of the dataset uncovers 268 noteworthy premiums [26] of the 3,282 premiums which seemed more than ten times those in the dataset. Grouping of the interests gave 42 unique topics. Names of topics were aggregated by the creators, and in this way have some subjectivity. Figure 10.4 demonstrates the most prominent and disliked topics in the medication group and the rest of the dataset. For each topic, we computed the likelihood of its event in the sedative or tranquilizer group and the rest of the dataset as a likelihood of the event of even one esteem from that subject.

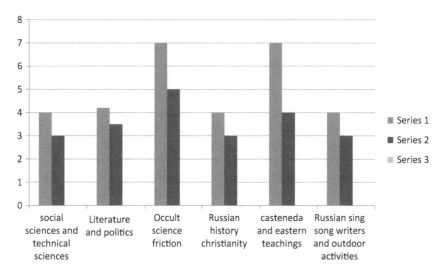

FIGURE 10.4
Characteristics of users writing about drugs.

10.5 Data Analytics

- **Phi(1):** Socioeconomic components deciding the position of life of the Territory in general, including the attributes of the stratification of society, change, the government-managed savings and different elements. Dynamic elements can be evaluated from the watched information and authority-measurable markers given by the administration [27]. Be that as it may, since official factual reports are utilized, the noteworthy dormancy of watched esteems ought to be considered. By and the large gathering of elements describes the general "condition" of the domain.

- **Phi(2):** Factors of the passionate and mental condition of the general public that are pointers of prosperity. These variables are normally dictated by sociological looks into and are not generally objective: for instance, exceedingly powerless to purposeful publicity.

- **Phi(3):** Affiliation of a person to a few classes of society. This factor is sensible because sociological investigates (LINK [1]) demonstrated that a few classes of society have more prominent defenselessness to tranquilizer utilization.

- **Phi(4):** Individual has interests like individuals who utilize drugs. Nearness of medication-related premiums can flag that client has a place with some subculture or groups that support sedative utilization of any kind. Data about individual interests and basic interests

TABLE 10.1

Persian Correlation between Age Factors for Group of People with Physical Drug Addiction

Age Group	Unemployment Level	Gini Index	Ratio of Morality and Fertility Rates	Ratio of Divorces and Marriages	Life Satisfaction
15–17	0.3692	−0.7866	0.7539	0.6892	0.6855
18–19	0.5143	−0.8771	0.8469	0.8501	0.6516
20–39	−0.3822	0.8188	−0.9837	−0.8513	−0.8438
40–59	−0.2032	0.6967	−0.7679	−0.8169	−0.8951

of individuals who utilize medications can be removed from online networking.

The model partitions each gathering—Phi(1), Phi(2), Phi(3), Phi(4)—into subgroups relying on age and sex of individuals.

Indicate size of every subgroup as Phi(n) (Table 10.1).

The condition of all age and sex subgroups in the gathering can be portrayed by the network whose general view is displayed at (10.1):

$$N^k = \begin{bmatrix} n_{1.1}^k & n_{2.1}^k \\ \cdots & \cdots \\ n_{1.100}^k & n_{2.200}^k \end{bmatrix} \tag{10.1}$$

$$A = \begin{bmatrix} 0 & 0 & \cdots & 0 & 0 \\ F_{.1}^{2i} & F_{.1}^2 & \text{.......} & 0 & 0 \\ 0 & 0 & \cdots & 0 & 0 \\ 0 & 0 & \text{.......} & F_{.i}^{2n-1} & 0 \end{bmatrix} \tag{10.2}$$

As noted before, each gathering portrayed framework changes after some time. Advancement of a can is depicted as far as Markov chain by network equation probabilities of the person with relating sex and age who can be categorized as one of five gatherings. Individual methodologies asymptotically take after three unique practices as indicated by the dynamic association of the people. Unadulterated dynamic clients are those people who are constantly dynamic all through the development procedure [28]. On the other hand, unadulterated turncoats are those people who continue prowling. A second-rate class is constituted by fluctuating people who on the other hand go about as dynamic clients or prowlers. The dynamic association of people partaking in the prowler round of an informal community appears in Figure 10.3. All people should be balanced ($\kappa = 0.1$).

As per the Matrix (5), when $\varepsilon < c/b$, dynamic clients dependably pick up a lesser result than prowlers. Companions' systems have no impact on the dynamic clients' choices. The dynamic clients will vanish in the system. The impact of clients gets more critical with the expansion in the estimation of ε. More clients will share data and result in the expanding division of unadulterated dynamic clients. At the point when ε is sufficiently vast, all the vertices in the system are instigated to end up unadulterated [29] dynamic clients by the impressive result.

Here, we talked about an extraordinary situation. An online informal organization is a substantial group, in which clients can get enthusiastic solace. It takes care of the relational correspondence requests and influences clients to feel near their companions.

At the point when a client peruses the remarks of his or her post, he or she feels self-satisfied. On the off chance that distributing a post gives a client passionate solace, it will end up noticeably as one of a client's propensities. The cost is about zero for the client, i.e., $c \to 0$ or $c \ll b$. At that point, the resulting lattice is:

$$\begin{array}{c} \\ A \\ L \end{array} \begin{array}{cc} A & L \\ \begin{bmatrix} E+b & eb \\ B & 0 \end{bmatrix} \end{array} \qquad (10.3)$$

Clients will pick the dynamic technique for the non-zero ε. The esteem c/b turns into the first purpose of ε. As appeared in Figure 10.4, the greater part of clients will wind up as noticeably unadulterated dynamic clients. Be that as it may, to build up the clients' propensity is as yet an incredible test for all the interpersonal interaction sites.

An extraordinary number of focal clients keep dynamic in a developed informal organization. These unadulterated dynamic clients constitute a focal bunch with high availability. The focal bunch assumes a huge part of the operation for entire system. Figure 10.5a demonstrates the topology for a system containing the information obtained from Weibo. The focal vertices are associated with each other to frame a focal group. In the meantime, each of them is associated with other peripheral vertices. Figure 10.5b demonstrates a schematic chart in which four dark focal vertices are associated with each other to frame a focal bunch. In the interim, each of them is associated with another two white negligible vertices. Focal vertices, for the most part, remain in the dynamic state. Because of their high inward network, the conditions of them can't be changed by the minor prowlers. In the entire system, few prowlers have no impact on the dynamic clients. These prowlers will be influenced by their dynamic neighbors and change their states (Figures 10.5 and 10.6).

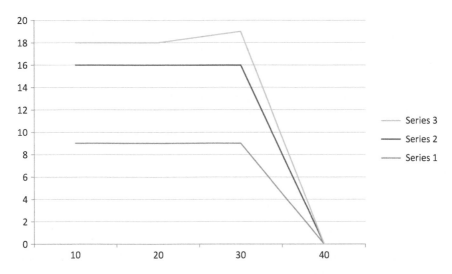

FIGURE 10.5
Fraction of pure fractuation strategies as a figure of G.

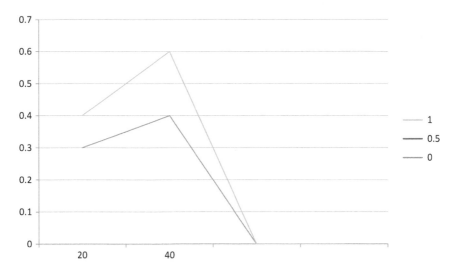

FIGURE 10.6
Fraction of pure and fluttering strategies as function of E.

The dynamic vertices develop amid developmental procedure by motiva-tion. As appeared here, 4,230 vertices have diverse procedure conveyances in various time, where $\varepsilon = 31$ and $\kappa = 0.2$. At first, the procedure is arbitrarily appropriated, and the focal vertices likewise embrace the hiding system. After a certain time, the vast majority of focal vertices will be modified to a dynamic state [30]. The prowlers, for the most part, exist in the vertices with medium and little degrees. The division of dynamic clients gets bigger with the expansion in the degree. At the point when the system is steady, the majority of vertices end up noticeably dynamic, aside from a few vertices with little degrees, which are associated with each other to frame some small bunches. These bunches are typically obliged in the greatly minimal areas and can scarcely speak with different parts of the system. They have a dull air and will progress toward becoming "perpetual prowlers."

In a system without impetus, all the vertices have inadequate motivating force. Dynamic focal vertices can scarcely influence the conditions of their neighbors and may even progress toward becoming prowlers because of the immense number of hiding neighbors. The previous discussion will help us to identify vertices with little degrees can keep dynamic because the little groups in the minimal regions shape a group segregated from different vertices. They have their advantages and points. Albeit no motivator is given, high grouping coefficient of these peripheral bunches can even now keep up a high action position.

An expansive number of systems demonstrate a propensity for the linkage arrangement between neighboring vertices. In other words, the system topology goes amiss from uncorrelated arbitrary systems, in which triangles are inadequate. This propensity is called bunching and mirrors the grouping of edges into firmly associated neighborhoods. The companions of a person are probably going to know each other.

$$C_i = 2t_i / k_i (k_i - 1) \qquad (10.4)$$

Where t_i signifies the number of triangles around I. Subsequently $C_i = 0$ if none of the neighbors of a client is associated, and $C_i = 1$ if the greater part of the neighbors is associated.

To consider the characteristic highlights of dynamic clients, we produce a simulated informal community with 42,000 clients as per past outcomes. The crossbreed arranges a model to portray the stresses of development and advancement of summed-up informal organizations. Of the key highlights, for example, a reasonable degree can be balanced physically. To disentangle the examination, there is just a single group in the produced organize. The process we propose will demonstrates the extent of dynamic clients versus the normal position of the system under three distinctive bunching coefficients. When the grouping coefficient is distinct esteem, the scope of dynamic clients is expanded with the expansion in the normal degree. It implies that the expansion of vertices in the system isn't impacted by the closeness. More companions and more substance rouse a current client to be more dynamic. At the point when the normal position of the entire system is positive esteem, the extent of dynamic clients is expanded with the expansion in the bunching coefficient. A client, for the most part, imparts normal enthusiasm to the companion. On the off chance that a client is occupied with a large portion of the posts, he or she is roused to be dynamic and may give remarks or forward the posts.

In a genuine informal community, a few clients are influenced by unreasonable elements, for example, singular contrasts, which result in practices against the amusement hypothesis. The previous graphs and plottings demonstrates the variety of thickness of dynamic clients with ε under various clamor conditions. At the point when people are almost discerning ($\kappa = 0.1$), the thickness of dynamic clients gets littler, and the pattern is the same as that appeared in Figure 10.3. At the point when the clamor is sufficiently expansive ($\kappa = 50$), everybody ends up noticeably nonsensical and does not pick the system as indicated by the amusement running the show. Every person in the system picks his procedure autonomously along these lines, prompting the vanishing of "hot locale." Extensive clamor produces the turmoil of systems. On the off chance that the constant disarray exists, dynamic clients are not ready to influence their neighbor prowlers, so they bit by bit lose interest and leave the system.

10.6 Completion and Analysis for Future Work

In this paper, we displayed our approach for mining and examination information from web sorting-out, which depends on mixing goliath information and passing on figuring flawless models. Guide Reduce demonstrate was utilized to mine, store and process monster measures of information from electronic frameworks organization. Utilizing Hadoop structure, we realized and appropriated a subject crawler for electronic frameworks organization. Getting ready for mined information is additionally performed by Hadoop, which improves the movement of new tallies and gives high flexibility and versatility. We proposed to utilize scattered figuring condition for streamed checks, which unravels the creation and execution of composite applications, which in turn perform complex information examination. The composite application works with information gotten from Hadoop by API and contains nuclear computational errands that accomplish some examination and showing objectives. We utilized the condition for passed-on, enrolling-based, cloud sort-out CLAIRE, which acknowledges AaaS show-up for spread preparing and gives predictable joining of isolated programming modules and that keeps running on various figuring assets. Along these lines this unequivocally streamlined plotting, executing, and reusing definitively finished applications.

Electronic frameworks organization contains a ton of individual data and can be utilized as an extra information hotspot for examination of social procedures in clear, particularly shapes that are not to any extent of the creative energy conspicuous in clear like the spread of disorders or interest and point of view of society toward sedative use. Despite the way that web sorting out can't supplant normal wellsprings of data that are accessible from social researchers or specialist real pointers as given by the organization, they can supplement this information. Another favored position of online social frameworks organization is that they give full-scale and small-scale qualities of clients. Immense-scale attributes are the general fervor of entire groups of clients toward some point, and more diminutive scale qualities are, for instance, centrality of the client to some social affair. Also, again web sorting-out uncovers some answers concerning more small-scale positions of social techniques while particular wellsprings of data are more about full-scale parameters.

We proposed to utilize information from web sorting-out as an extra wellspring of data for examination and showing of medication use among the majority of a certain zone. Examination of the small-scale position of clients who expound on drugs enabled finding out about their interests, which distinguish them from others. We got their mental picture, which incorporates subjects of interest that give off an impression of being powerfully (or less) routine in this social event. We utilized the vocabulary of arrangement-related catchphrases and key explanations to pick whether the client writes in blog areas about medications. We assume that among various clients who have specific motivations to clean up illicit drug use, we discovered some strategies of clients who share their experience of utilizing drugs. This is circuitously confirmed by the closeness of such concentrations of terms like "Russian shake," "non-common cure," "astounding," "eastern lessons," and which, particularity in Russia, are all-around considered medicine-related. Regardless, to affirm this, further research is required. The depicted approach utilizes a broad assortment of parameters and breaking points which on an essential position impact on the quick overview of basic interests. Regardless, our trials showed that rundown of monstrous themes remains predictable. We portrayed the model for the guess of different individuals that have unmistakable positions of solution obsession in the predefined zone. We gauged tons of the full-scale and small-scale condition of the general populace on a foreordained zone. Two or three full-scale parts can be surveyed in light of the official information sources; however, a position basic to drug use is stunningly less hard to assess in context of the information from web sorting-out. Furthermore, official information sources cover individuals with solid physical solution dependence; however, electronic frameworks organization spreads individuals with light medication fixation profiles. We displayed estimations of parts respects for solid ward individuals yet estimation of elements in light of the information from web sorting-out and more appearing within these segments is future work.

In any case, some feedback about utilizing web sorting-out ought to be noted; for instance, not all general populace parties are shown additionally in online social frameworks organization, whose essential clients are adolescents. Likewise, there are responses of faithfulness [22] and unfaltering quality [23] of utilizing information from online frameworks organization which looks to a great degree sensible. In spite of these reactions, we accept that the examination of social affiliations is basic and can uncover data about frameworks that are disguised or can't be especially seen in people in the general field. This work was monetarily kept up by the Government of the Russian Federation, Grant074-U01.

10.7 Conclusion

We implemented a better approach with data mining and big data analytics with cloud computing. We manage data with the data mining approaches to capture and handle data, and, using the advanced MapReduce approach, we started implementing a novel architecture which will handle all the types of data, and the cloud computing is used as the repository, and the implementation is very simple as mentioned in the proposed system approach. The architecture contains all the important dependencies like MapReduce, data mining technology and the web-based implementations to handle the data like data inflow, management, and outflow. The application utilizations are our architecture in this model. In future analysis we would like to implement a better machine learning model including fog computing, which will create a separate network implementations for every application on the web.

Acknowledgment

We would like to extend our sincere thanks for the researchers who worked on the previous versions of the research presented in this article. The help provided by the other scholars from different organizations is most valuable for us and we thank them from the bottom of our hearts. The research done previously by different scholars helped us very much to learn new things from their work. Thanks for those technocrats.

References

1. A. Yakushev, A. Boukhanovsky, P.A. Sloot, Topic crawler for social networks monitoring, P. Klinov, D. Mouromtsev (Eds.), *Knowl. Eng. Cement. Web SE - 17*, Springer, Berlin, Germany, pp. 214–227, 2013.

2. X. Sun, B. Gao, Y. Zhang, W. An, H. Cao, C. Guo et al., Towards delivering analytical solutions in cloud: Business models and technical challenges, *E-Business Engineering (ICEBE), 2011 IEEE 8th International Conference on*, pp. 347–351, 2011.

3. D.J. Abadi, Data management in the cloud: Limitations and opportunities, *IEEE Data Eng. Bull.*, vol. 32, pp. 3–12, 2009.

4. A. Kittur, B. Suh, E.H. Chi, Can you ever trust a wiki?: Impacting perceived trustworthiness in Wikipedia, *Proceedings of the 2008 ACM Conference on Computer Supported Cooperative Work*, pp. 477–480, 2008. https://dl.acm.org/citation.cfm?doid=1460563.1460639.

5. J. Mangeksdorf, Supercomputing the climate: NASA's Big Data Mission, *Comput. Inf. Sci. Technol. Off.*, 2012.

6. V. Kadam, S. Tamane, V. Solanki. Smart and connected cities through technologies, in *Big Data Analytics for Smart and Connected Cities*, IGI-Global, doi:10.4018/978-1-5225-6207-8

7. A.J. Jara, Y. Sun, H. Song, R. Bie, D. Genooud, and Y. Bocchi, Internet of Things for the cultural heritage of smart cities and smart regions, in *Proceedings of theIEEE 29th International Conference on Advanced Information Networking and Applications Workshops (WAINA)*, pp. 668–675, March 2015.

8. A.T. Campbell et al., The rise of people-centric sensing, *IEEE Internet Comput.*, vol. 12, no. 4, pp. 12–21, 2008.

9. D.D. Sanju, A. Subramani, and V.K. Solanki, Smart city: IoT based prototype for parking monitoring & parking management system commanded by mobile app, in *Second International Conference on Research in Intelligent and Computing in Engineering*.

10. R. Dhall and V.K. Solanki, An IoT based predictive connected car maintenance approach, *IJIMAI*. doi:10.9781/ijimai.2017.433.

11. I. Butun, M. Erol-Kantarci, B. Kantarci, and H. Song, Cloud-centric multi-position authentication as a service for secure public safety device networks, *IEEE Commun. Mag.*, vol. 54, no. 4, 2016.

12. C. Jiang, Y. Chen, and K.J.R. Liu, Graphical evolutionary game for information diffusion over social networks, *IEEE J. Sel. Topics Signal Process.*, vol. 8, no. 4, pp. 524–536, 2014.

13. V.K. Solanki, M. Venkatesan, and S. Katiyar, Conceptual Model for Smart Cities for Irrigation and Highway Lamps using IoT, *IJIMAI*. doi:10.9781/ijimai.2017.435.

14. R. Gross and A. Acquisti. Information revelation and privacy in online social networks. In *Proceedings of the 2005 ACM Workshop on Privacy in the Electronic Society, WPES'05*, pp. 71–80, New York, ACM, 2005.

15. V.K. Solanki, M. Venkatesan, and S. Katiyar, Think home: A smart home as digital ecosystem in circuits and systems, vol. 10, no. 7.

16. K. Lee, J. Caverlee, and S. Webb, Uncovering social spammers: Social honeypots + machine learning. In *Proceedings of the 33rd International ACM SIGIR Conference on Research and Development in Information Retrieval, SIGIR'10*, pp. 435–442, 2010.

17. V.K. Solanki, S. Katiyar, V.B. Semwal, P. Dewan, M. Venkatesan, and N. Dey, Advance automated module for smart and secure city, in *First International Conference on Information Security and Privacy (ICISP 2015)*, G.H. Raisoni College of Engineering & Information Technology, Nagpur, India, December 11–12, Procedia Computer Science, Elsevier, 2015.

18. P. Domingos, Mining social networks for viral marketing, *IEEE Intell. Syst.*, vol. 20, pp. 80–82, 2005.
19. J. Bollen, H. Mao, and X. Zeng, Twitter mood predicts the stock market, *J. Comput. Sci.*, vol. 2, pp. 1–8, 2011.
20. D.J. Abadi, Data management in the cloud: Limitations and opportunities. *IEEE Data Eng. Bull.*, vol. 32, pp. 3–12, 2009.
21. M. Diao, S. Mukherjea, N. Rajput, and K. Srivastava, Faceted search and browsing of audio content on spoken web, in *Proceedings of the 19th ACM International Conference on Information and Knowledge Management*, pp. 1029–1038, 2010.
22. D. Dash, J. Rao, N. Megiddo, A. Ailamaki, and G. Lohman, Dynamic faceted search for discovery-driven analysis, in *ACM International Conference on Information and Knowledge Management*, pp. 3–12, 2008.
23. T. Cheng, X. Yan, and K.C.-C. Chang, Supporting entity search: A large-scale prototype search engine, in *Proceedings of the ACM SIGMOD International Conference on Management of Data*, pp. 1144–1146, 2007.
24. M. Bron, K. Balog, and M. de Rijke, Ranking related entities: Components and analyses, in *Proceedings of the ACM International Conference on Information and Knowledge Management*, pp. 1079–1088, 2010.
25. W. Dakka and P.G. Ipeirotis, Automatic extraction of useful facet hierarchies from text databases, in *Proceedings of the IEEE 24th International Conference on Data Engineering*, pp. 466–475, 2008.
26. A. Herdagdelen, M. Ciaramita, D. Mahler, M. Holmqvist, K. Hall, S. Riezler, and E. Alfonseca, Generalized syntactic and semantic models of query reformulation, in *Proceedings of the 33rd International ACM SIGIR Conference on Research and Development in Information Retrieval*, pp. 283–290, 2010.
27. M. Mitra, A. Singhal, and C. Buckley, Improving automatic query expansion, in *Proceedings of the 21st Annual International ACM SIGIR Conference on Research and Development in Information Retrieval*, pp. 206–214, 1998.
28. P. Anick, Using terminological feedback for web search refinement: A log-based study, in *Proceedings of the 26th Annual International ACM SIGIR Conference on Research and Development in Information Retrieval*, pp. 88–95, 2003.
29. X. Xue and W.B. Croft, Modeling reformulation using query distributions, *ACM Trans. Inf. Syst.*, vol. 31, no. 2, p. 6, 2013.
30. J. Pound, S. Paparizos, and P. Tsaparas, Facet discovery for structured web search: A query-log mining approach, in *Proceedings of the ACM SIGMOD International Conference on Management of Data*, pp. 169–180, 2011.

11

IoT Recommender System: A Recommender System Based on Sensors from the Internet of Things for Points of Interest

Cristian González García, Daniel Meana-Llorián, Vicente García Díaz, and Edward Rolando Núñez-Valdez

CONTENTS

11.1 Introduction

Currently, different services use in their background recommender systems to give a specific and custom-made recommendation for each user. This is the case of AliExpress, Amazon, HBO, Netflix, or even Google and Facebook when showing us advertisements. This is possible due to the use of a specific recommender system that analyzes all the data that we produce or the data from people who have similar tastes to ours. Using these data, recommender systems can suggest to us customized contents.

Nevertheless, there are different types of recommender systems [1]. Some of them need specific content by users, like reviews or scored items to create personalized recommendations. Other times, they use metadata, like the things that were searched, visited, or read. Notwithstanding, the former is something that users do not usually do. Then, when we need to recommend different points of interest (POIs), we need this information and in this case, this creates a lack of data to do the customized recommendations. However, with the Internet of Things (IoT) [2–4], this situation could change drastically because of the heterogeneous and ubiquitous sensors that could be dispersed around the world [5,6], allowing us to use any object to do anything [7–9].

To solve this situation, in this article we propose a hybrid system where the information is retrieved through explicit feedback. This recommender system will use the feedback from users and from sensors to give different recommendations for the user who does a query to the system, as Cha et al. demonstrate [10]. However, in our case, we want to extend all the work to any area, not only tourism, and including personal data about similar users to improve the system.

Using the idea that we present here, it will be possible to obtain feedback from similar users who have been in places and have given a "like" to those places; the different sensors that are distributed in different places allow us to gather data from the exact POI that we want to evaluate. Clearly, in these cases, sensors will have to be registered under the same POI or to be placed closely, inside or around the POI, which will be evaluated to use the exact valid sensor to create good recommendations.

Notwithstanding, the major problem in this case is how to analyze all the large amount of data, what method or methods to use to do that, and how to create correct recommendations. This proposal presents the main idea of IoT recommend systems and how the problems could be solved with different solutions to create a recommend system that uses the sensors of the IoT.

The remainder of this work is structured as follows: Section 11.2 presents the state of the art, which contains the related work and what technologies will be used in this proposal. Section 11.3 shows the proposal and the general ideas and solutions to fit and solve our goals. Section 11.4 contains the conclusions. Finally, Section 11.5 presents future work to be done.

11.2 State of the Art

In this section, we introduce some of the most important concepts to understand the proposal. We start with the Internet of Things.

11.2.1 Internet of Things

The Internet of Things (IoT) is one of the most important topics in research and business [2,3,11,12]. Its goal is the interconnection between heterogeneous and ubiquitous objects among themselves. It was considered as one of the six technologies with most interest to the EEUU until 2025 by the United States National Intelligence Agency [13].

As we already said, the aim of IoT is the interconnection of heterogeneous and ubiquitous objects and different systems among one another. A requirement to achieve that interconnection is the capacity for the objects and systems to connect to the Internet [14]; so, IoT exists to extend the Internet to things [15].

The term "Internet of Things" emerged in a presentation of Kevin Ashton for Procter & Gamble (P&G) about the use of the technology Radio-frequency identification (RFID) in supply chains [16] in 1999. He introduced the word "Internet" because it was a trending topic and he wanted to attract attention. However, he never defined this term, and of course, he did not expect its future impact.

In our proposal, we want to connect distributed sensors around different places, with users getting recommendations relative to their preferences and the data collected by sensors in real time. These sensors are objects in the Internet of Things. In the next subsection, we are going to introduce the objects of the Internet of Things: the smart objects and the non-smart objects.

11.2.2 Smart Objects, Sensors, and Actuators

An important piece of the Internet of Things is the smart objects and the not-smart objects. Smart objects or intelligent products [17] are physical elements capable of interacting with everything around them, the environment and other objects. Their behavior depends on the data that they process, and it can be automatic or semi-automatic. They are usually capable of reacting according to interactions with other smart objects [3,18]. We can consider smart TVs, smart phones, tablets, some cars, etc. as examples of this type of object. Smart objects can be classified in three dimensions [6], which represent qualities of their intelligence: level of intelligence, or how much intelligence an object can have; location of intelligence, or where the intelligence is located; and aggregation level of intelligence.

However, not-smart objects also exist. These objects lack intelligence [6]. They usually compose other objects or smart objects. Whereas smart objects can work without any dependency, non-smart objects are devices that need another device to work. Sensors and Actuators are two examples of non-smart objects. They need another object that can process the data that sensors measure, like temperature, or, in the case of actuators, they need another object that can order them the actions to do according to certain conditions.

11.2.3 Recommender Systems

Recommender systems are intelligent systems that use mechanisms and techniques applied to information retrieval. These systems are very important tools that help users to reduce the search time of content on the web that can be of interest for them, efficiently and effortlessly. These systems help users to find any kind of content, such as books, movies, electronic products, websites, songs, hotels, insurance, etc. in a relatively easy way for users [19].

Recommender systems tries to solve the information overload problem on the web and facilitate the retrieval of information, through the implementation of algorithms and information classification mechanisms [20–22].

According to the information-filtering paradigm that is used, in general, recommender systems can be classified in several approaches, which include **collaborative filtering, content-based**, and **hybrid approach**. Other authors also propose other classifications of recommender systems, such as **utility-based, knowledge-based**, and **demographic** [23].

Recommender systems collect information related to users' profiles using **explicit** or **implicit feedback techniques**. This information is used to provide valid information to users. For example, in Donovan and Smyth [19], the authors present an approach where a recommender system can help users in finding electronic books in a social network using a mixed feedback techniques (explicit and implicit). This mixed approach represents another paradigm for recommender systems.

Many e-commerce websites and online social networks like Amazon store, AliExpress, Facebook, LinkedIn, and other types of websites such as Film affinity, Netflix, and HBO, have a recommender system to offer interesting content to its users.

Finally, in the IoT world, these kinds of intelligent systems can help discover points of interest (beaches, hotels, parks, etc.) using sensor networks available on the Internet. The integration of recommender systems with a real-time sensor network can help minimize travel time and costs, and increase users' satisfaction.

11.2.4 Artificial Intelligence

Artificial intelligence (AI) is a concept that refers to the intelligence exhibited by machines or intelligent agents, i.e., any device that perceives its environment and performs actions that maximize its chance of success from the point of view of some specific goal [24]. AI is a huge and interdisciplinary field founded on the claim that human intelligence can be so precisely described that a machine could be created to mimic it [25]. Thus, AI is based on a range of different disciplines such as computer science, biology, psychology, linguistics, mathematics, or sociology.

To achieve its goal, AI usually relies on different approaches. For example, search and optimization algorithms are important for reasoning and leading from premises to conclusions. Fuzzy logic is a version of first-order logic which allows the truth of a statement to be represented as a value between 0 and 1, rather than just 1 or 0. Probabilistic methods (e.g., Bayesian networks, Hidden Markov model, etc.) are very interesting for uncertain reasoning. Classifiers and statistical learning methods such as machine learning give computers the ability to learn without being explicitly programmed. Neural networks are computational models used to solve problems in the same way that the human brain would do.

There are plenty of areas in which AI can be applied. For example, Devedzic [26] worked on a survey of web-based education applications in which adaptability and intelligence is important, through the use of semantic web technologies. KL-ONE is a well-known system that has been used in both basic research and implemented knowledge-bases systems for representing knowledge [27]. Others, like Bennett and Hauser [28], worked on developing a general-purpose framework to explore various healthcare policies and payment methodologies and to be the basis for clinical artificial intelligence.

One classic application of AI is to support recommendation systems. They are a clear example of intelligent systems. For example, Veena and Babu [29] propose a user-based recommendation system addressing challenges in collaborative filtering and scalability based on Apache Mahout. Lam et al. [30] addressed the cold-start problem in recommendation systems using hybrid approaches, with the analysis of two probabilistic aspect models that put together collaborative filtering and information from users. There are also many more general works that deal with issues related to machine learning [31].

Another current and rising use of AI is the Internet of Things. For example, Gubbi et al. [32] present a cloud-centric vision for worldwide implementation of the Internet of Things. They discuss the key enabling technologies and application domains in which machine learning methods are an essential aspect. Among the many related works, we can highlight some, such as Khan et al. [33], that explain the architecture, applications and key challenges on the Internet of Things, in which machine-to-machine communication is also essential to embed some form of intelligence.

The Internet of Things, recommendation systems, and AI can also work together. An interesting work is the one by Von Reischach et al. [34]. They present an idea to enable consumers to access and share product recommendations using their mobile phone by using RFID codes to receive and submit product rankings. Mashal et al. [35] work on a platform to recommend services in IoT (e.g., personal care, energy monitoring) by a graph-based formal model. There are some authors that work on topics more related to our proposal. One very interesting work is the one by Sun et al. [36]. They work on smart communities by connecting devices and smart sensors. They propose the use of personal sensors, open data, and sensing services in tourism and cultural heritage with a context-aware recommendation system. Cha et al. [10]

propose a platform for supporting a real-time recommendation system based on IoT smart connected devices. They show their approach with a prototype of a tourism-based application to demonstrate the entire process.

11.3 Proposal

In this section, we are going to introduce our proposal. Our goal is to give recommendations about POIs to users which will be based on data recollected by sensors from the IoT and previous users' experiences. Throughout this section, we will address how we will collect data from users and from sensors, how the system will access a list of places to recommend, and the different solutions to process the collected data. Our proposal focuses on the creation of a novel recommendation system that users will be able to use to know relevant places that fit a query made by them. For accomplish this goal, we propose the creation of a novel system that will gather relevant data to users' queries from sensors distributed around different places, and it will allow users to give feedback about the given recommendations. Thus, we will have two sources of information: the users themselves and the distributed sensors. Throughout both sources of information, we will obtain a score per place according to users' preferences and sensors' data. Moreover, we will give more recommendations based on similar preferences of other users. Hence, we are proposing a hybrid recommendation system (content-based and also based on collaborative filtering) where the information is retrieved through explicit feedback.

This section is composed by four different subsections: (1) obtaining information from the users themselves and from sensors; (2) the registration of places; (3) the interaction of users with the proposed system; and (4) the different possible implementations of the score calculator whose goal is to obtain a score by each place in an intelligent way.

11.3.1 Feedback of Information

Our proposal consists in the creation of a novel recommendation system that will suggest different POIs to users according to their preferences and conditions of the places through distributed sensors. Thus, we will have two different sources of feedback, the users and the sensors.

11.3.1.1 Feedback from Users

Our proposal is a hybrid recommendation system whose information is retrieved through explicit feedback. First, we are going to explain why the feedback will be explicit and after, why the recommendation system will be hybrid.

The proposal will give several recommendations to users according to their queries, allowing them to visualize the information and indicate whether they like such recommendations. By this way, we obtain what their theoretical preferences are. Moreover, users will have the possibility of rating the place when they are there. Due to that, we will obtain a rate about a place with specific conditions in real time. So, we will collect feedback before the user goes to the place and when the user is there. By this way, we will know the preferences of each user and his or her level of satisfaction with each recommendation, taking into consideration the real conditions which will be obtained through the sensors.

11.3.1.2 Feedback from Sensors

Our proposal will collect data from sensors distributed around the available places. These set of sensors will be sensors registered in our system by the users themselves or public sensors available in other services through public application programming interfaces (APIs). Our goal is the creation of a recommendation system for places, hence, we will collect data only from some types of sensors.

The registration of sensors in our system is an important issue to address. Our system will be able to use sensors from third-party services and sensors registered by our users. Thus, we will have to enable users to register their sensors and make the system compatible with third-party services.

How users will able to register their sensors will be addressed in a next section, when we will address how users will be able to interact with the proposed system. How to use sensors from third-party services is also very important. We propose the creation of different categories of sensors related to environment conditions like temperature, humidity, air quality, wind speed, and so on. These conditions will be used to classify each place according to the different values. Thus, each place will have a temperature, a humidity, etc.

When a user makes a query, like for the best beaches of the north of Spain, the system will collect all registered places and it will retrieve the actual status though the sensors registered in the system and make requests to the registered third-party services through their APIs. Therefore, the system will have a list of places with their actual conditions. To make good recommendations, the system will consult the preferences of the users and the POIs they liked, since they should have given a good value from previous visits. From this information, the system will select a set of POIs and calculate a score per place to give suggestion to users.

11.3.2 Registration of Places

Our proposal will need a list of POIs which will be recommended to users. To make this list, we identified three possibilities to register places: (1) manually; (2) automatically on demand; and (3) predictive and automated.

- **Manually:** With the purpose of registering POIs that our proposal can use, we propose that users could register places manually through a form in the web application. The principal advantage of this approach is that places will be already registered in our system when a user makes a query. Nevertheless, the number of registered places would be too limited to satisfy any query of every user. Moreover, it could show errors related with the introduction of wrong information during the manual registration.

- **Automatically on demand:** On the other side, the information on POIs would be collected from external services without the need of manual interaction, hence, the registration of places could be automatic. An external service could be Google Places; nevertheless, we cannot process all places available in this service. Therefore, the process of collecting information would begin when users make queries, on demand. A case of use could be the following: a user makes a query about beaches in the north of Spain, then, the system will search all beaches in the north of Spain in Google Places and compare the results with the places stored in our proposal. Nevertheless, if these places were not registered in the system, they would be registered to future queries.

- **Predictive and automated:** Finally, the last approach that we suggest could be a predictive and automated one. Although we are introducing this approach as another option, it would be complementary to the automatic one because it is based on automatically collecting places that users will search through Google Places. The principal issue is the identification of possible future searches. An example of how to address it could be the following: if users made many queries about places of the same location, the system would identify this location as a potential one for future queries, hence, the system would register automatically more places of this location from Google Places.

11.3.3 Users Interface

In this chapter, we propose the creation of a novel recommendation system that suggests places according to a query made by users and their personal preferences. These preferences will be registered in the system through valuing the recommendations. When the system recommends several places, users will be able to choose what recommendations they like, and moreover, when users will be in a recommended place, they will be able to rate this place. However, the first time that users will use our proposal, they will not have any previous interaction that enables the system to know their preferences. Therefore, when users will be registered, they will have to fill in a form with their preferences to help the system to make the first recommendations.

To make queries, users will have to use a web application with a form designed for this purpose. Throughout this form, the system will know what users will want. The queries would be composed of a location, a distance, type of place, and other filters that will enable the system to do accurate recommendations.

After making the query, users will be able to choose the recommendations that they would like to save and consult them later, because they will be able to assign value to these recommendations before going to the places and after being there. In this way, the system will know where users were and make more recommendations according to users that usually go to same places with similar conditions.

11.3.4 Different Solutions for the Score Calculator

Here, we have different solutions to create the score calculator: two methods for the feedback from users, and two possibilities for the feedback from sensors. These methods are the following:

The first one is related to the **feedback from users**. Here, we can search similar users with a similar taste for the POI that the query user wants. This could be possible using a comparator between the profiles of each user to obtain the most similar users. In this way, the recommender system could offer places that would maybe match the query made by the user based on similar users' tastes. Here, to give a score to each profile, we have two possibilities: One of these possibilities is using the old method of comparing one field with another field to obtain the profiles with more coincidences. Another possibility is applying knowledge discovery databases (KDD) [37] to obtain valid patterns about what type of places, usually liked by one type of user, which must have been categorized previously in the same type of profile. For instance, a tanning-beach lover who likes the beach without wind, a very sunny day and without an opinion about the waves of the sea; a surfer-lover who likes the high waves but does not care about the sun, or a 7-a-side football lover who probably likes cloudy days without wind, and so on.

In the second case, to obtain the **feedback from sensors**, we could create an algorithm, using Fuzzy Logic [38], to mix each linguistic variable and obtain a final score of that profile. In this case, we will have to create adaptive controllers [39], also known as expert systems [40], and membership functions [41]. These controllers have the rules that are used to process the linguistic variable in the fuzzification process [42]. After the whole process, we will obtain the final score after doing the defuzzification from the last fuzzy number to the final score that will have the POI interest for the query made by the user.

On the other hand, we can customize the adaptive controllers and the membership functions for each user in the case of using a mechanism to train, learn, and evolve in time for each user. It could be possible to create it if we base the rules on different information to personalize the adaptive controllers and the membership functions for each user. This system will be an adaptive fuzzy logic system [43]. In this case, we will have to use some machine-learning technologies like Bayesian networks, artificial neural networks, decision trees, association rule learning, support vector machines, clustering, reinforcement learning, representation learning, genetic algorithms, or deep learning [37].

11.4 Conclusions

In this proposal, we have presented an idea that defines a new recommender system by improving the recommendations using the information of sensors of the IoT. This novel idea could give recommendations based on the exact weather conditions for the taste of each user. This is so because sometimes the same place has different weather conditions and not all of us like the same place when it's raining or when it's being sunny. It depends on what type of place it is, like a beach.

Then, this IoT recommender system could allow us to include in recommender systems a new external variable, which is the weather conditions like the quantity of light, the speed of the wind, if it is raining or not and how much, the waves of the sea, the crowd, and so on. These data with the preferences of users could be able to give recommendations about different POIs according to the exact weather and the different tastes of people. Then, this idea is something that could improve the current recommendations.

Therefore, we could improve the recommender systems using personal tastes and the current condition of some POIs. Nevertheless, the different possible solutions should be studied and compared in search of the best solution to well fit the system.

11.5 Future Work

In order to accomplish this recommender system, we will need to apply different branches of the computer science fields, but it will not be totally finished due to the many different ways in which this work could be continued. We show some of these future ways as follows:

- **Big data infrastructure:** To analyze in near-real time the different queries and to create good recommendations, as well as the storing of all the data, we will have to research an optimal infrastructure which could use different big data tools to store, process, manage and execute all the data [37].

- **Comparison among the different machine learning methods:** As we have explained before, we could use different machine learning methods to create an adaptive fuzzy logic system. Notwithstanding, we will not know which of these methods could produce the best results. This is why we will need to do a comparison among all of them.

- **Include more metadata to the recommendations:** The recommendations always could be improved. In this way, other metadata then could be included to improve the IoT recommender system. For example, it could use different data from smart phones in real time, like the GPS or the places that the user is visiting, to try to predict the next place they are going to visit based on the user route of that day.

Acknowledgments

This work was performed by the "Ingeniería Dirigida por Modelos MDE-RG" research group at the University of Oviedo under Contract No. FC-15-GRUPIN14-084 of the research project "Ingeniería Dirigida Por Modelos MDE-RG." The project was financed by PR Proyecto Plan Regional.

References

1. E.R. Núñez-Valdez, Sistemas de Recomendación de Contenidos para Libros Inteligentes, University of Oviedo, 2012.
2. C. González García, C.P. García-Bustelo, J.P. Espada, G. Cueva-Fernandez, Midgar: Generation of heterogeneous objects interconnecting applications. A domain specific language proposal for Internet of Things scenarios, *Comput. Networks.* 64 (2014) 143–158. doi:10.1016/j.comnet.2014.02.010.
3. L. Atzori, A. Iera, G. Morabito, The Internet of Things: A survey, *Comput. Networks.* 54 (2010) 2787–2805. doi:10.1016/j.comnet.2010.05.010.
4. C. González García, J.P. Espada, Using model-driven architecture principles to generate applications based on interconnecting smart objects and sensors, in: V.G. Díaz, J.M.C. Lovelle, B.C.P. García-Bustelo, O.S. Martinez (Eds.), *Advances and Applications in Model-Driven Engineering*, IGI Global, 2014: pp. 73–87. doi:10.4018/978-1-4666-4494-6.ch004.

5. C. González García, J.P. Espada, MUSPEL: Generation of applications to interconnect heterogeneous objects using model-driven engineering, in: V.G. Díaz, J.M.C. Lovelle, B.C.P. García-Bustelo (Eds.), *Handbook of Research on Innovations in Systems and Software Engineering*, IGI Global, 2015: pp. 365–385. doi:10.4018/978-1-4666-6359-6.ch15.

6. C. González García, D. Meana-Llorián, B.C.P. G-Bustelo, J.M.C. Lovelle, A review about smart objects, sensors, and actuators, *Int. J. Interact. Multimed. Artif. Intell.* 4 (2017) 7–10. doi:10.9781/ijimai.2017.431.

7. D. Meana-Llorián, C. González García, B.C.P. G-Bustelo, J.M. Cueva Lovelle, N. Garcia-Fernandez, IoFClime: The fuzzy logic and the Internet of Things to control indoor temperature regarding the outdoor ambient conditions, *Futur. Gener. Comput. Syst.* (2016). doi:10.1016/j.future.2016.11.020.

8. C. González García, D. Meana-Llorián, B.C. Pelayo G-Bustelo, J.M. Cueva Lovelle, N. Garcia-Fernandez, Midgar: Detection of people through computer vision in the Internet of Things scenarios to improve the security in Smart Cities, Smart Towns, and Smart Homes, *Futur. Gener. Comput. Syst.* (2017). doi:10.1016/j.future.2016.12.033.

9. D. Meana-Llorián, C. González García, B.C. Pelayo G-Bustelo, J.M. Cueva Lovelle, V.H. Medina García, IntelliSenses: Sintiendo Internet de las Cosas, in: *2016 11th Iberian Conference on Information Systems and Technologies (CISTI)*, Gran Canaria, Spain, 2016: pp. 234–239. doi:10.1109/CISTI.2016.7521551.

10. S. Cha, M.P. Ruiz, M. Wachowicz, L.H. Tran, H. Cao, I. Maduako, The role of an IoT platform in the design of real-time recommender systems, in: *Internet Things (WF-IoT), 2016 IEEE 3rd World Forum*, 2016: pp. 448–453.

11. K. Gama, L. Touseau, D. Donsez, Combining heterogeneous service technologies for building an Internet of Things middleware, *Comput. Commun.* 35 (2012) 405–417. doi:10.1016/j.comcom.2011.11.003.

12. C. González García, J.P. Espada, E.R.N. Valdez, V. García-Díaz, Midgar: Domain-specific language to generate smart objects for an Internet of Things platform, in: *The 8th Intertnational Conference on Innovative Mobile and Internet Services in Ubiquitous Computing*, IEEE, Birmingham, UK, 2014: pp. 352–357. doi:10.1109/IMIS.2014.48.

13. The US National Intelligence Council, Six Technologies with Potential Impacts on US Interests out to 2025, 2008.

14. G.M. Lee, J.Y. Kim, Ubiquitous networking application: Energy saving using smart objects in a home, in: *2012 International Conference on ICT Convergence*, IEEE, Jeju Island, 2012: pp. 299–300. doi:10.1109/ICTC.2012.6386844.

15. S. Li, L. Da Xu, S. Zhao, The internet of things: A survey, *Inf. Syst. Front.* 17 (2014) 243–259. doi:10.1007/s10796-014-9492-7.

16. K. Ashton, That "Internet of Things" thing, *RFiD J.* 22 (2009) 97–114. http://www.itrco.jp/libraries/RFIDjournal-That Internet of Things Thing.pdf.

17. G.G. Meyer, K. Främling, J. Holmström, Intelligent products: A survey, *Comput. Ind.* 60 (2009) 137–148. doi:10.1016/j.compind.2008.12.005.

18. C.Y. Wong, D. McFarlane, A. Ahmad Zaharudin, V. Agarwal, The intelligent product driven supply chain, in: *IEEE International Conference on Systems, Man, and Cybernetics*, IEEE, 2002: p. 6. doi:10.1109/ICSMC.2002.1173319.

19. E.R. Núñez-Valdez, J.M.C. Lovelle, G.I. Hernández, A.J. Fuente, J. Labra-Gayo, Creating recommendations on electronic books: A collaborative learning implicit approach, *Comput. Human Behav.* (2015). doi:10.1016/j.chb.2014.10.057.

20. J. O'Donovan, B. Smyth, Trust in recommender systems, in: *Proceedings of the 10th International Conference on Intelligent User Interfaces*, ACM, New York, 2005: pp. 167–174. doi:10.1145/1040830.1040870.
21. P. Resnick, N. Iacovou, M. Suchak, P. Bergstrom, J. Riedl, GroupLens: An open architecture for collaborative filtering of netnews, in: *Proceedings of the 1994 ACM Conference on Computer Supported Cooperative Work*, ACM, New York, 1994: pp. 175–186. doi:10.1145/192844.192905.
22. P.O. de Pablos, R.G. Crespo, O.S. Martínez, J.E.L. Gayo, B.C.P. García-Bustelo, J.M.C. Lovelle, Recommendation System based on user interaction data applied to intelligent electronic books, *Comput. Hum. Behav.*, 27 (2011) 1445–1449. doi:10.1016/j.chb.2010.09.012.
23. G. Adomavicius, A. Tuzhilin, Toward the next generation of recommender systems: A survey of the state-of-the-art and possible Extensions, *IEEE Trans. Knowl. Data Eng.* 17 (2005) 734–749. doi:10.1109/TKDE.2005.99.
24. S. Russell, P. Norvig, *Artificial Intelligence: A Modern Approach*, Vol. 25. Prentice Hall, Egnlewood Cliffs, NJ (1995).
25. M. Minsky, *The Emotion Machine: Commonsense Thinking, Artificial Intelligence, and the Future of the Human Mind*, Simon & Schuster, 2007.
26. V. Devedzic, Education and the semantic web, *Int. J. Artif. Intell. Educ.* 14 (2004) 39–65. http://iospress.metapress.com/index/HR4V08QM6VY8Y3T7.pdf (accessed December 24, 2014).
27. R.J. Brachman, J.G. Schmolze, An overview of the KL-ONE: Knowledge representation system, *Cogn. Sci.* 9 (1985) 171–216. doi:10.1016/S0364-0213(85)80014-8.
28. C.C. Bennett, K. Hauser, Artificial intelligence framework for simulating clinical decision-making: A Markov decision process approach, *Artif. Intell. Med.* 57 (2013) 9–19.
29. C. Veena, B.V. Babu, A user-based recommendation with a scalable machine learning tool, *Int. J. Electr. Comput. Eng.* 5 (2015).
30. X.N. Lam, T. Vu, T.D. Le, A.D. Duong, Addressing Cold-start Problem in Recommendation Systems, in: *Proceedings of the 2nd International Conference on Ubiquitous Information Management and Communication*, ACM, New York, 2008: pp. 208–211. doi:10.1145/1352793.1352837.
31. F. Ricci, L. Rokach, B. Shapira, *Introduction to Recommender Systems Handbook*, Springer, 2011.
32. J. Gubbi, R. Buyya, S. Marusic, M. Palaniswami, Internet of Things (IoT): A vision, architectural elements, and future directions, *Futur. Gener. Comput. Syst.* 29 (2013) 1645–1660. doi:10.1016/j.future.2013.01.010.
33. R. Khan, S.U. Khan, R. Zaheer, S. Khan, Future internet: The internet of things architecture, possible applications and key challenges, in: *Frontiers of Information Technology (FIT), 2012 10th International Conference on*, 2012: pp. 257–260.
34. F. Von Reischach, D. Guinard, F. Michahelles, E. Fleisch, A mobile product recommendation system interacting with tagged products, in: *Pervasive Computing and Communications 2009. PerCom 2009. IEEE International Conference on*, 2009: pp. 1–6.
35. I. Mashal, T.-Y. Chung, O. Alsaryrah, Toward service recommendation in Internet of Things, in: *Ubiquitous and Future Networks (ICUFN), 2015 Seventh International Conference on*, 2015: pp. 328–331.

36. Y. Sun, H. Song, A.J. Jara, R. Bie, Internet of things and big data analytics for smart and connected communities, *IEEE Access.* 4 (2016) 766–773.

37. C. González García, MIDGAR: Interoperabilidad de objetos en el marco de Internet de las Cosas mediante el uso de Ingeniería Dirigida por Modelos, University of Oviedo, 2017. doi:10.13140/RG.2.2.26332.59529.

38. L.A. Zadeh, Fuzzy logic and approximate reasoning, *Synthese.* 30 (1975) 407–428. doi:10.1007/BF00485052.

39. P. Grant, A new approach to diabetic control: Fuzzy logic and insulin pump technology, *Med. Eng. Phys.* 29 (2007) 824–827. doi:10.1016/j.medengphy.2006.08.014.

40. J.R. Stuart, N. Peter, *Artificial Intelligence: A Modern Approach*, 2nd ed., Prentice Hall, Upper Saddle River, NJ, 2003.

41. L.A. Zadeh, Fuzzy sets, *Inf. Control.* 8 (1965) 338–353. doi:10.1016/S0019-9958(65)90241-X.

42. E. Portmann, A. Andrushevich, R. Kistler, A. Klapproth, Prometheus—Fuzzy information retrieval for semantic homes and environments, in: *3rd International Conference on Human System Interaction*, IEEE, Rzeszow, Poland, 2010: pp. 757–762. doi:10.1109/HSI.2010.5514482.

43. W.A. Kwong, K.M. Passino, Dynamically focused fuzzy learning control, *IEEE Trans. Syst. Man Cybern. Part B.* 26 (1996) 53–74. doi:10.1109/3477.484438.

12

Internet of Things: A Progressive Case Study

Aditee Mattoo and Somesh Kumar

CONTENTS

12.1 Introduction

The Internet of Things (IoT), a growing field in the technology industry, also referred to as Internet of Everything (IoE) or Industrial Internet of Things (IIoT), is envisioned as a global network of machines that aims at interconnecting network-enabled devices. The concept of Internet-connected devices evolved in early 1982, with the first appliance being a Coke machine at Carnegie Mellon University that was able to report its inventory and verify that newly loaded drinks were cold or not. Its popularity began to rise in 1989 through the technology radio-frequency identification (RFID), coined by Kevin Ashton, who preferred the phrase Internet for Things instead of Internet of Things [1]. RFID systems consist of readers and smart low-cost electronic tags that are responsible for carrying out automatic identification of the object and assign a unique digital identity to each object to be integrated in the network. RFID acts as an electronic barcode that helps in an automatic identification of devices. The IoT aims at providing interconnection between objects to create a smart environment by achieving smart connectivity and context-aware computation through three things [2]:

- A mutual understanding of the state of its users and their physical devices
- Software design architectures and persistent communication networks to process and deliver the contextual or textual information to where it is relevant
- The analytics and logical tools in the Internet of Things that aim for autonomous, intelligent, and smart behavior

The main idea and concept behind a smart environment is to make machines smart enough to almost nullify human effort. The vision of IoT is to develop interconnected devices where these appliances share information with humans, cloud-based applications, and to each other as in device-to-device, smartly and efficiently.

The IoT can be defined as the network that connects any living/nonliving object which exists in the real-time world and can be monitored through the Internet. The existing trend in IoT can provide an interconnection of objects and things for establishing smart environments. Three essential components in developing a smart product are its physical infrastructure, smart behavior, and connectivity. Physical infrastructure consists of hardware circuit realization, electrical configuration, etc.; smart components include sensors, microprocessors, and analytics; and the third component, connectivity, includes ports and connection devices that send the data from the product up to the cloud. To attain a smart environment, unique identification of objects is important for favorable outcomes of the IoT. Along with unique identification and reliability, scalability and persistence are the key requirements to develop an IoT-based product. Figure 12.1 depicts IoT architecture. Wireless communication is established through an IoT framework by using embedded hardware that transmits and detects the data through sensors like temperature, pressure, lux, humidity, etc. [3].

Key objectives of smart product are as follows:

1. **Monitoring:** Smart products' functional characteristics are monitored by companies and customers to keep track of them. These products enable the intensive and thorough monitoring of a product's condition, operation, and external environment through sensors and external data sources. Using monitored data, a smart product can alert or notify changes in circumstances or performance to users or others. Monitoring is most helpful in the case of medical devices.

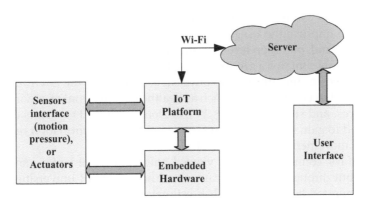

FIGURE 12.1
IoT architecture.

2. **Controlling:** Smart, connected products can be controlled through remote commands or algorithms that are built into the device or reside in the product cloud. Algorithms are rules that direct the product to respond to specific changes in its condition or environment.

3. **Optimization:** Smart, connected products can apply algorithms and analytics to in-use or historical data to dramatically improve output, utilization, and efficiency. In wind turbines, for instance, a local microcontroller can adjust each blade on every revolution to capture maximum wind energy. Real-time monitoring data on product condition and product control capability enable firms to optimize service by performing preventative maintenance when failure is imminent and accomplishing repairs remotely, thereby reducing product downtime and the need to dispatch repair personnel.

4. **Automation:** Monitoring, control, and optimization capabilities combine to enable smart, connected products to achieve a previously unattainable level of autonomy. Sophisticated products are able to learn about their environment, self-diagnose their own service needs, and adapt to users' preferences. Autonomous products can also act in coordination with other products and systems. The value of these capabilities can grow exponentially as more and more products become connected.

Smart devices are designed such that they capture and utilize every bit of data that has been shared in everyday life to interact with the users on a daily basis and complete tasks that have been programmed. Other IoT-related technologies include wireless sensor networks (WSNs), Wi-Fi, and embedded sensor and actuator nodes that transform the current static Internet into an integrated future Internet. The IoT utilizes the Internet to incorporate heterogeneous devices with one another. In order to achieve the target, sensors can be used at different locations for collecting and analyzing data to improve the usage. Various environmental conditions like temperature, sound, pollution levels, humidity, wind, etc. can be measured with WSNs. A WSN contains a group of spatially dedicated and dispersed sensors that are used for monitoring the physical conditions of the environment. It may be used in various sectors such as healthcare, government, environmental services, and seismic sensing. WSNs along with RFID systems could be integrated to gain some output in terms of obtaining information regarding position, movement, temperature, etc. [4,5].

Nowadays, connectivity is not viewed only in terms of laptops and smart phones, but through connected cars, smart homes, smart universities, smart cities, connected healthcare, etc. It is estimated that there will be a huge rise in the number of connected devices, and everything will be moved toward an automated world. IoT devices bridge the gap between the physical and the digital world to improve the overall scenario of life, society, and industries. IoT platforms are suites of software components that work as middleware between hardware and application layers. They help to integrate sensors, devices, networks, and software to work together and develop a smart product.

12.2 IoT and Different Application Scenarios

IoT is an emerging technology that is creating a lot of buzz throughout the world. The IoT utilizes the Internet as a network to incorporate heterogeneous devices with one another. Different sensors are used at different locations for collecting and analyzing data that improves the usage of devices. Various applications of the IoT are nowadays emerging as smart devices that are transforming our lives. The IoT is creating its impact on the entire world in form of smart homes, smart cities, wearable devices, connected cars, and in agricultural and healthcare domains. Some of these applications are given in detail below [6–10]:

1. **Smart Homes:** An IoT platform at home enables the automation of common activities. Smart home applications use sensor networks for monitoring by using the data generated by the sensors. Basically, smart homes are the ones in which various appliances are wired to a central computer control system so that they can be either switched on or off at certain times. Most homes already have smart features, as they contain built-in sensors or electronic controllers. A smart home IoT platform constantly monitors the state of the home, light levels, and switches appliances to be on and off accordingly. It can also detect movements across the floor and respond to the events. It can also switch on light and music in different rooms on detection of movement.

2. **Smart Cities:** Smart cities are one of the promising applications throughout the world. A smart city can include smart surveillance, smarter energy management systems, water distribution, automated transportation, urban security, and environmental monitoring. Smart cities' efficiency can be increased by focusing on solving problems faced by the people in the city, like pollution, vehicular traffic congestion, energy supplies shortage, etc. With the help of sensors, smart parking lots can be established across the city that can track arrival and departure of various vehicles. Weather and water systems can also be controlled by analyzing the information like temperature, pressure, wind speed, and rain. Energy distribution and consumption in heterogeneous circumstances can also be managed by optimal scheduling of energy suppliers. In emergency situations, fault detection, service restoration, and isolating, for example, are handled by the application by determining the position of the defective parts.

3. **Wearables:** This application has a great influence in market all over the world. Various companies like Google, Apple, and Samsung have invested heavily in building these devices. For these wearable applications, prerequisite are energy-efficient, ultra-low, and small-sized devices. These wearable devices collect data and information about the users with the medium of sensors and software installed in the devices. The data gathered is preprocessed and essential information about the user is extracted. The device contains and covers fitness, health, and entertainment essentials.

4. **Connected Cars:** Large automakers as well as start-ups are working on connected car solutions to enhance the in-car experience. These vehicles will be able to optimize their own operations, maintenance, and real-time data acquisition and will ensure comfort of passengers by using onboard sensors and Internet connectivity.

5. **Industrial IoT (IIoT):** The IIoT uses sensors, softwares, and big data analytics for creating brilliant machines that are more accurate and consistent through data communication. This helps in monitoring the data, picking problems and inefficiencies in a well-defined approach to rectify it. The IIoT is also beneficial for quality control and can increase the supply-chain efficiency by tracking goods and providing real-time information exchange about inventory among suppliers and retailers with automated delivery.

6. **Agriculture IoT:** As the world population is increasing, there is a continuously rising demand on the food supply. Governments are helping farmers to use advanced technology and increase food production by using meaningful insights from the data to yield a better return on investment. The features of smart farming include sensing for soil moisture, nutrients, controlling water usage for plant growth, and determining custom fertilizers.

7. **Healthcare IoT:** In healthcare, the IoT has a massive importance. It will aim at empowering people to live healthy lives by using smart connected devices. The data generated from the devices will help in personalized analysis of an individual's health. Patient monitoring, personnel monitoring, disease spread modeling and containment, real-time health status and predictive information to assist practitioners in the field, or policy decisions in pandemic scenarios are used as facilities in smart health.

12.3 Connectivity and Languages Supported by IoT

For establishing connectivity, the first stage is identifying data sources. After that methods of acquiring data from those sources should be identified. Coding in the particular software has to be done to access sensor readings and translate those readings into human-understandable form. Also, these softwares can help in reading log file data and parsing that log data file into properties. After acquisition of data, data has to be moved in concert with the IoT cloud. After that embedded designing, connectivity with the network, and final user interface is developed. Connectivity can be wired or wireless [11].

12.3.1 Wireless Architecture

In wireless architecture, various methods are described as follows:

1. **Adapter Cellular:** This technique includes connecting a device to a cell tower, accessing its network, and connecting to the Internet. Smart devices use this technology by installing a cellular communication module like in mobile devices, vehicles, and in equipment that move around frequently. Cellular networks can be GSM, LTE, CDMA, WLAN, etc.

2. **Adapter Wi-Fi:** Connectivity is achieved through a gateway modem, a hardware device that connects the device to the Internet. Other wireless technologies are in the form of 6LowPan, Sub-1GHz, Bluetooth, and Zigbee. These are all variations of wireless networking.

12.3.2 Wired Architecture

1. **Adapter Ethernet:** Wired architecture is also used by devices for transmitting a lot of data. Ethernet connectivity is generally used with desktop computers, printers, commercial devices, and industrial or consumer devices, such as a laptop. For the case where device doesn't

have any built-in communication facilities, the device is hardwired to an external controller that adds both programmable automation logic and connectivity and runs in a stand-alone mode. An example of this is a programmable logic controller (PLC) connected to a device such as a hydraulic press or an injection molding machine.

2. **PLC:** It is connected to a control computer that automates a single hardwired device, which is then used to automate the device as part of larger establishments. It directs PLC using industrial communications protocols such as MQTT, MODBUS, Allen Bradley, etc. as it loops through a logical set of instructions. After the looping process, PLC then executes its commands on the hardwired device. To enable connections from the device network to the Internet, a gateway is inserted into this network, which is responsible for transporting all the necessary data from the device network to the cloud and interfacing with the control computer.

Various softwares support IoT device programming that are primarily responsible for monitoring and controlling the device. The device acts as an agent that reads data from data sources like sensors, actuators, or log files running on the machine where software is installed. Some software can help in changing the device configuration, updating the drivers and firmware, or even optimizing its operating procedures.

12.4 Language Overview—Arduino and LUA

12.4.1 Arduino

Arduino is an open source–based electronic programmable board consisting of hardware (microcontroller) and software as an integrated development environment (IDE). For loading a program into a controller board, no extra hardware is needed. Instructions are passed to the microcontroller, with help of which any activity can be performed. It accepts both analog and digital signals as input and produces desired output. Arduino boards are used to read inputs, like lights on a sensor or a finger on a button, and generate an output such as activating a motor, turning on/off an LED, or online data publishing. Arduino is used in different projects and applications because of its simple and accessible user experience. The various types of microcontroller supported by Arduino are as follows [12]:

- ATMEGA328 microcontroller
- ATMEGA3244 microcontroller
- ATMEGA2560 microcontroller

- ATMEGA168 microcontroller
- AT91SAM3 × 8E microcontroller

Arduino consists of microcontroller ATmega328P, operating voltage 5V, input voltage from 7–12 V that is recommended, input voltage (limit) 6–20 V, 14 digital input/output pins, 6 PWM digital input/output pins, 6 analog input pins, DC current per input/output pin 20 mA, DC current for 3.3 V pin 50 mA, flash memory 32 KB (ATmega328P) of which 0.5 KB is used by boot loader.

Various advantages of using Arduino board are as follows [13–15]:

- **Economical:** Arduino boards are inexpensive as compared to other microcontroller platforms. The Arduino board can even be accumulated by hand, and even the pre-assembled Arduino modules are economical.

- **Cross-platform:** The Arduino IDE runs on Windows, Macintosh OSX, and Linux operating systems, and most microcontroller systems run on only Windows.

- **Simple, clear programming environment:** The Arduino software is easy to use for beginners as well as flexible for advanced users. Various inbuilt programs are provided to give an overview of the way Arduino works.

- **Open source and extensible software:** The Arduino software has open source tools that can be extended by experienced programmers. C++ libraries are used for expanding language, and AVR-C programming language is used for understanding the technical details of the board. Also, this AVR-C code can be directly added into Arduino programs.

- **Open source and extensible hardware:** The outline of the Arduino boards are published under a creative common license. Experienced circuit designers can extend, improve, and transform the old version into the customized version. Inexperienced users can build the breadboard version of the module in order to understand the module better and also save money.

Arduino IDE is an open source software platform that is used to program the Arduino controller board. The programming of the board is based on the deviations of C and C++ programming language. Just like C programming, Arduino also supports various operators like arithmetic, comparison, boolean, bitwise, and compound. The steps required to upload the code are as follows:

- Connect USB to computer.
- Upload program in chip, launch Arduino IDE.
- Select board type and port.

A program that is written into the Arduino board is called a sketch. There are two main parts of a sketch: setup () and loop (). Setup () is written at the beginning of the code, where it is used to initialize input-output variables and pin modes. It is written once when reset is pressed or power is provided to board.

```
void setup(){
Serial.begin(9600);}
```

In the code above, setup function is used to begin serial communication; baud rate (9600) indicates bits of data traveled per second between board and computer. For establishing communication between an Arduino board and a computer or other devices, the term "serial" is used. All Arduino boards have at least one serial port (also known as a UART or USART). The serial port communicates on digital pins 0 (RX, i.e., receive) and 1 (TX, i.e., transmit) and with the computer via USB. Thus, if these pins are used as a serial port, then they can't be used for digital input or output. For communication with an Arduino board, the Arduino environment's built-in serial monitor can be used. For using that feature, click the serial monitor button in the toolbar and select the same baud rate used in the call to begin(). Serial communication on pins TX/RX uses TTL (transistor–transistor logic) logic levels (5 V or 3.3 V depending on the board). These pins should not be directly connected to an RS232 serial port, as they operate at ±12 V and can damage the Arduino board.

Arduino Mega has three additional serial ports: Serial1 on pins 19 (RX) and 18 (TX), Serial2 on pins 17 (RX) and 16 (TX), and Serial3 on pins 15 (RX) and 14 (TX). These pins can communicate with the personal computer, by using an additional USB-to-serial adaptor, as they are not connected to the Mega's USB-to-serial adaptor. For using these ports to communicate with an external TTL serial device, connect the TX pin to the device's RX pin, the RX to the device's TX pin, and the ground of the Mega to the device's ground.

The Arduino DUE has three additional 3.3V TTL serial ports: Serial1 on pins 19 (RX) and 18 (TX); Serial2 on pins 17 (RX) and 16 (TX), and Serial3 on pins 15 (RX) and 14 (TX). Pins 0 and 1 are also connected to the corresponding pins of the ATmega16U2 USB-to-TTL serial chip, which is connected to the USB debug port. Additionally, there is a native USB-serial port on the SAM3X chip, called SerialUSB.

The Arduino Leonardo board uses Serial1 to communicate via TTL (5V) serial on pins 0 (RX) and 1 (TX). Serial is reserved for USB CDC (communication device class) communication.

Other function is loop (), that runs over and over again forever.

```
void loop(){
Serial.println("Hello board");
delay(1000);
}
```

The above piece of code prints Hello board with delay of 1000 milliseconds in between reads for stability.

The following example shows the blinking of the LED on the Arduino board.

```
// the setup function runs once when you press reset or power
the board
       void setup() {
// initialize digital pin 13 as an output.
       pinMode(13, OUTPUT);
       }

// the loop function runs over and over again forever
       void loop() {
       digitalWrite(13, HIGH); // turn the LED on (HIGH is the
       voltage level)
       delay(1000); // wait for a second
       digitalWrite(13, LOW); // turn the LED off by making
       the voltage LOW
       delay(1000); } // wait for a second
```

12.5 LUA Scripting Language

Interpreted languages, known as glue languages, are split into two applications: kernel and a configuration written in two different languages. The kernel part implements the core components of the system and is written in a complied language like C or C++. In the configuration part, an interpreted language connects all the components to give a final design to the application [16].

Lua, a Portuguese word meaning "the moon," is a programming software that is especially designed for embedded systems. It is simple to process as server statements are available, embedded in C, C++, Java, FORTRAN, etc. where Wi-Fi connection is simply built. The language is portable and integrates well with other platforms with dynamic structures. It's safe environment, open source availability, automatic memory management, and virtuous facilities for handling strings and other kinds of data with dynamic size make it efficient and reliable. Lua is an independent and interpreted language that allows precompiled parts of program to run on different machines and handles communication between the components written in C. CPU (central processing unit) tasks are handled by C functions. The other feature of Lua comprises of execution of chunks of code that has been created dynamically. Lua is said to be an event-driven programming model where each message is taken as an event, involving a simple task such as invoking a remote function and a complex task such as remotely changing an algorithm that a process is running. Features of Lua are as follows [17]:

- **Embeddability:** Lua is written in ANSI C and has a relatively simple C API with low-level operations. It supports bidirectional features; i.e., host calls Lua and vice versa.

- **Portability:** Runs on different machines like Unix, MacOS, Windows, PS2, PS3, etc. and has no direct dependency on the operating system. Lua also supports embedded hardware.

- **Simplicity:** Lua is a simple language with one data structure, i.e., tables that can represent arrays, sets, lists, and records, and are essentially heterogeneous associative arrays.

- **Small Size:** Lua is a lightweight language with an entire distribution of 209 KB. The reference interpreter is only about 180 KB compiled, and it is easily adaptable to a broad range of applications.

A lightweight programming language is designed to have very small memory footprint, is easy to implement (important when porting a language), and/or has minimalist syntax and features. In computing, the memory footprint of an executable program indicates its runtime memory requirements, while the program executes. This includes all sorts of active memory regions like code segments containing (mostly) program instructions (and occasionally constants), data segments (both initialized and uninitialized), heap memory, and call stack, plus memory required to hold any additional data structures, such as symbol tables, debugging data structures, open files, shared libraries mapped to the current process, etc., that the program always needs while executing and will be loaded at least once during the entire run. NodeMCU is designed to run on all ESP modules and includes general purpose interface modules which require at most two GPIO pins. Lua script is flashed into the ESP8266 processor and compiled on the suitable port. When the connection is made, the code runs through it and displays the output.

NodeMCU firmware implements Lua 5.1 over Espressif SDK for its ESP8266 SoC (system on chip), and the IoT modules are based on this. NodeMCU Lua is based on eLua that is a featured implementation of Lua 5.1 and is best suitable for embedded system development and its execution. NodeMCU Lua provides a scripting framework that is used for useful applications within the limited RAM and flash memory resources of embedded processors like that of ESP8266. The NodeMCU development kit is used to open the appropriate COM (communication) port for the ESP8266 processor through USB cable and a SPIFFS (serial peripheral interface flash file system) that requires a good Wi-Fi connection. To enable ESP8266 firmware flashing, GPIO0 pin must be pulled low before the device is reset. Table 12.1 shows the operations and values required for communication between the ESP8266

TABLE 12.1

Establishing Communication between
ESP8266 and Port

Operation	Value
COM PORT	COM5
AP MAC	1A-FE-34-CA-D6-07
STA MAC	18-FE-34-CA-D6-07

and system using NodeMCU firmware flasher [18,19]. These values are subjected to change with respect to the machine used.

After establishing communication, init.lua script is uploaded in the ESPlorer application. Any modification in the script can be done in a real-time basis. Port is opened up and script is saved and run on the ESPlorer. After the final script is ready, it is saved to ESP8266 and a connection is established on a particular IP address. Once the script is saved onto the ESP8266 processor, it remains there in the memory and no need to connect to the USB every time. Lua script can be modified in real time as per the requirement of the project. Example code that shows how to control and set output mode is as follows:

```
// Controlling LED settings & output mode
    pin_led = 7
    pin_inblt = 4
    pin_bulb = 2
    gpio.mode(pin_led, gpio. OUTPUT)
    gpio.mode(pin_inblt, gpio. OUTPUT)
    gpio.mode(pin_bulb, gpio. OUTPUT)
    gpio.write(pin_led, gpio. LOW)
    gpio.write(pin_inblt, gpio. HIGH)
    gpio.write(pin_bulb, gpio. HIGH)
```

Here, firstly input variables (pin_led, pin_inblt, and pin_bulb) are initialized with the pin numbers of ESP8266 chip. After that mode of gpio (general purpose input output). pins are set as input or output. Then these pins are set to their default condition as low or high according to the requirement.

12.6 Case Study and Implementation: Automatic Light System

Embedded system is a computer system having an amalgamation of a dedicated function comprising of a larger electrical system with real-time computing parameters; i.e., a combination of hardware and software. The software is embedded into computer hardware read-only memory (ROM) and does not need secondary memory for storage. Embedded processors have low power consumption, small size, and low per unit cost of materials, which makes them efficient to use. These systems are assigned to perform specific tasks rather than general purpose tasks and are useful in automatic appliances, industrial applications, medical sciences, security, wireless technology, commercial, military, and transportation units. The appliances that are making use of embedded systems are mobile phones, microwaves, washing machines, GPS, cabs, etc. Apart from the above mentioned appliances, there are customized systems that offer high performance at lower costs and saves energy, hence reducing manual efforts. The computing systems can range from having no user interface—that is, the systems that perform only single tasks like buttons, touchscreens, LEDs and sensors—to the complex ones having graphical user interface like in mobile applications. There are two basic types of embedded systems [20,21]:

1. **Embedded system hardware:** This is the most commonly used system that is designed to perform real-time operations. System hardware can be microcontroller or microprocessor based and both containing integrated circuits. The basic difference between both the types is that the microprocessor implements only the CPU, and other components like memory chips can be added as self-contained systems, whereas microcontrollers implement CPU, memory and other peripherals or ports. This system hardware communicates with both microcontroller and computer. Universal serial bus (USB) is used as the interface to communicate with the computer, and the Internet service provider (ISP) to communicate with microcontroller.

2. **Embedded system software:** This system uses operating systems or a specific language platform and is applicable for real-time operating environments. Embedded systems can be stand-alone or real-time systems with network and mobile devices. It performs well-defined tasks fast with accuracy and efficiency. Along with microprocessor and microcontrollers, it is comprised of protocol converters, communication interfaces, and drivers. The quality of embedded software is analyzed by checking the communication among the interfaces, devices, and ports.

The objective of this system is to design an embedded system for automatically controlling the light system. Hardware circuit and programmable interface are important for the realization of the system. A low-cost, economical, and efficient system has to be used so that it can be applied in any organization, thus saving power. Various input sensors have to be properly utilized that can they convert the analog signals to digital ones. All the information is passed to the microcontroller, which can save the information in memory and produce analog output that users can see in LCD displays or in the form of audio accessible through speakers. The illumination system developed should be reliable, efficient, and economical.

Therefore, interfaces are designed using the Lua programming environment, and communication of the server is established with sensory ports and other sub-units. Each controller will have sensors like LDR (light-dependent resistor), current, voltage, temperature, and a real-time clock to detect the suitable light pattern and take action as desired by the device at that time. Small, simple, and low-cost microcontrollers are used that provide greater capability, efficiency, and accuracy in dealing with the system. The various steps involved in realizing the circuit are as follows:

- Firstly, analyze the design of Wi-Fi–based microcontroller server system.
- Review various algorithms and techniques used for developing the automatic light system.
- Design and fabricate a prototype model for establishing real-time RISC (reduced instruction set computing) -based wireless communication of the processor with sensors and sub-components for automatic light system application.
- Integrate the circuit designed with Lua and end-user Android programming interface using Firebase authentication and database on the basis of different controlled parameters and constraints.
- Analyze the complete automatic light system and verify with the authorized end users.

Thus, an embedded system is designed with hardware and a software application that generates an automatic light system, which in turn can save light and power in any university, institution, organization, or large establishments effectively and efficiently.

"Automatic light system architecture" gives the overall description of the major components used for remote communication of the RISC server-based processor with sensory ports and other sub-units through the Lua interface, controlled by a PDA (personal digital assistant) device. Various segments are incorporated together so that the communication among the peripherals may be achieved. The major components used in the system are the RISC processor, Wi-Fi modem, hardware module, Lua interface, and web server; along with the components, a graphical user interface is created so that end users can modify or view the current pattern. The output—i.e., a smart light system—is generated that provides the different light patterns on the basis of time, area, and distance between the poles of light for the university, thus saving light and human energy [22–25]. A basic block diagram of the system is depicted in Figure 12.2.

12.6.1 Hardware Configuration of Enabled Products

A Hardware platform is developed that allows the components to fit into the system architecture and provide communication between the ports like

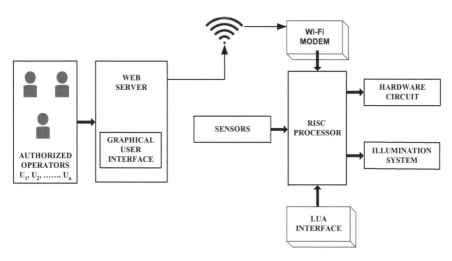

FIGURE 12.2
Block diagram of the light system.

Bluetooth, MIC, sensors, devices, processors, and programmable interface. Sensors like LDR, current, voltage, temperature, and a real-time clock are used that feed input to the processor. LDR works on the principle of photoconductivity, which means that when there is light available, the LDR will be in the OFF state, and when it becomes dark outside, the LDR will be in the ON state. When light falls on an LDR, that sensor sends commands to the microcontroller, depicting that the LDR should be in the OFF state. All commands are received by the microcontroller, based on which the device operates. A microcontroller is a small computer on a single integrated circuit, a system on a chip model that contains one or more processors with memory and programmable input or output peripherals. The RISC microprocessor is used in a project that executes a small, optimized set of instructions. The web server that is the main server carries all the important and updated information about the university's lights, poles, distance between the poles, area of illumination covered by the light, intensity of lamps, etc. [26,27].

The ESP8266 was designed and fabricated in China by Espressif Systems. Espressif have also developed and released a companion software development kit (SDK) to enable developers to build practical IoT applications for the ESP8266. The SDK is made freely available to developers in the form of binary libraries and SDK documentation.

ESP8266 Wi-Fi Wemos D1 mini, a microprocessor, is the integral component of the system and is used in the circuit designing. It is a Wi-Fi module that has a self-contained SOC (system on chip) that integrates all the components of a circuit into a single chip, along with a TCP/IP stack that gives the microcontroller access to Wi-Fi. The ESP8266 processor is a compact and small-size processor because of its power amplifier, low-noise receive amplifier, and power management modules that minimize PCB (printed circuit board) size by reducing the external circuitries. High-speed cache present inside the processor helps in increasing the system performance and optimizes the system memory. It also works as a Wi-Fi adaptor through interfaces like SPI (serial peripheral interface bus), secure digital input output, and UART (universal asynchronous receiver/transmitter) for serial communication. The chip integrates directly with Lua language and sensors through GPIOs (general purpose input output pins) as it contains a 32-bit Tensillica MCU. It is used in home automation, smart plugs, light, sensor networks, security cameras, etc. Some important protocols and features of ESP8266 are as follows [28,29]:

- Wi-Fi Direct (P2P Support)
- 802.11 b/g/n/e/i support
- WPA/WPA2 PSK/Pre-authentication security
- Frequency range between 2.4 GHz~2.5 GHz
- Tensillica 32-bit microcontroller
- Operating voltage of 2.5~3.6 V
- Operating current of average value of 80 Ma
- AES encryption
- IPv4, TCP/UDP/HTTP/FTP network protocols
- Micro USB connection

A microprocessor ESP8266 has a direct Wi-Fi (P2P support) and has direct communication with the Lua interface. It also consists of a Tensillica 16-bit RISC and 32-bit microcontroller unit. With its Wi-Fi supportability, it can connect to the Internet from any place and provides a medium for smart phones to interact with the system. Technical specifications of ESP8266 include Tensillica 32-bit microcontroller unit and ultra-low-power 16-bit RISC that consists of eight digital pins, from D1 to D8 and analog pin A0, as shown in Table 12.2.

Apart from the main component—that is, the Tensillica microcontroller discussed above—the hardware circuit includes various other essential parts that take input, produce output, detect and emit light, maintain current flow, regulate voltage and connectors, etc. Table 12.3 presents the list of the hardware components used in the realization of the circuit.

12.6.2 Fabrication Process

The hardware circuit for automatic light control system is designed by using EAGLE 7.2.0 Professional tool. EAGLE is easily applicable graphical layout editor and is best suited to design and realize hardware circuits. It is the powerful, easy-to-use, and freeware tool for any engineer or technically sound

TABLE 12.2

ESP8266 Pin Specifications and Related Functions

Pin Name	ESP8266 PIN	Function
3V3	3.3 V	Supply voltage
G	GND	Ground
D0	GPIO16	Input/output
D1	GPIO5	Input/output with SCL (synchronous clock line)
D2	GPIO4	Input/output, SDA (synchronous data line)
D3	GPIO0	Input/output,10 K pull-down
D4	GPIO2	Input/output, built-in LED
D5	GPIO14	Input/output, SCK (serial clock)
D6	GPIO12	Input/Output, MISO (master in slave out)
D7	GPIO13	Input/output, MOSI (master out slave in)
D8	GPIO15	Input/output, 10 K pull down
A0	A0	Analog input, 3.3 V
RX	RXD	Receive signal
TX	TXD	Transmit signal
RST	RST	Reset
5 V	–	5 V

TABLE 12.3

List of Components and Their Configuration

Component	Configuration
LDR sensor	Light-dependent resistor sensor
PRESET	Variable resistor 3296
LED sensor	Light-emitting diode sensor
OptoCoupler (OC)	MOC3041 (627Q)
TRIAC	BT139
Temperature sensor	LM35
Voltage regulator	LM7805
Resistors	330 Ω,360 Ω,470 Ω,39 Ω/1 W,1M Ω
Capacitors	104 J/400 V, 47 Nk, 103 (0.01 μF), 1000 μF
Diodes	IN4007

person to make schematic capture and printed circuit board (PCB) design for a circuit or combination of circuits. EAGLE displays fast designing with high-speed technology like DDR4 (double data rate), which is a fourth-generation synchronous dynamic random-access memory. The EAGLE professional tool is an open source and freeware tool; it can be installed on the system having an operating environment of Windows 7 or new versions with supportability of a 64-bit OS. It can also be used on the Linux and Mac systems and supports the minimum graphic version of 1024 × 768 pixels. Various features of the EAGLE tool are as follows:

- **Modular Design Blocks:** The blocks or modules of circuitry can be reused in different conditions.
- **Multi sheet Schematics:** Helps in keeping and maintaining the designs of any size organized and in order.
- **Electrical Rule Checking (ERC):** Once the circuit schematic is done, the designer can verify and check for errors from ERC.
- **Design Rule Checking (DRC):** DRC is used to test the PCB layout for any missed air wire connections, overlaps, etc., and to check whether the boards are ready to build. These rules are also valid for Auto router.
- **Real-Time Design Synchronization:** The tool provides a reliable feature of connection and sync between schematics and PCB. If both the designs are synchronized, then the final layout could never be proper and up to mark.
- **Simplified Selection and Editing:** Selecting, editing, and grouping the PCB design is a comfortable task.

12.6.3 Working of EAGLE

EAGLE tool works in two stages to correctly build and synchronize the circuit design. The two stages are the schematic and the PCB design to make the circuit ready to be built on board. In the first stage, schematic capture is prepared. The circuit connectivity is designed by looking for the proper components, their size, their type and grid; hence adding them to the editor and giving them proper connections through the net. Each component's name and value are added from the Name and Value tag. Move, Rotate, and Copy features are used to systematically adjust the component at the right position. Text can be written over any component, as required in the design. If the connections are getting complicated and less space is between the pins of the IC (integrated circuit), then labels can be added simply by giving similar names to both the specific pins of components to be connected. Junctions can be added between two net wires. The schematic design is saved as a .sch file.

Once the schematic is designed, EAGLE itself asks to create a PCB layout of the design and saves the PCB file with a .brd extension. All the components and their net wires are already placed in the PCB layout. The main job of PCB design is to correctly route the air wire connections from one component to the other and to verify that both schematic and PCB design are in synchronization or not. Figure 12.3 shows the schematic circuit for an automatic light control system where different ICs are used. The ESP8266 processor takes analog inputs from sensors LM35, LDR, etc. and provides output in the form of LED. The voltage regulator regulates the power supply with the help of TRIAC and opto-coupler.

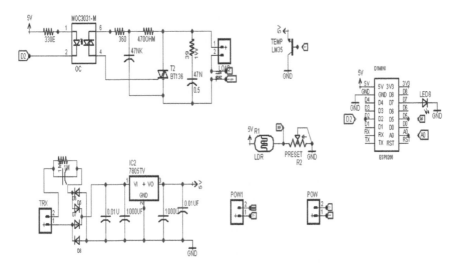

FIGURE 12.3
Schematic design for automatic light-control system.

The circuit shows how the components are connected together and how they work. It consists of an ESP module on whose digital pins D7, D2, and D5 are connected to LED, OC that is connected to TRIAC, and LDR respectively. Analog input is provided by theLM35 temperature sensor connected to A0, analog pin of ESP8266. IC 7805 regulates voltage and therefore the power circuit is used so that each circuit gets the proper voltage supply. A step-down transformer is used with value 9-0-9 (18V) that steps down 220VAC to 9V with maximum of 500 mAmp current. Output is produced in the form of the glowing of LED, lamp, and inbuilt LED of the module.

In the second stage, components are positioned on the PCB and then the routing between the components takes place. The net connections are placed as they are designed in the schematics, and with help of them, routing is done. The auto-routing feature is also available in PCB designing but manual routing is preferred as it helps to provide the liberty to connect the components in accordance with the designer and moreover, extra layers can be avoided. Various layers like top, bottom, pads, unrouted, and dimensions are present and can be selected or unselected as per the requirement. For a reliable and efficient design, top layers should be avoided, otherwise, on board in actual implementation, manual connection through wires has to be given. Routing could be corrected again if gotten wrong, with the help of rip-up tool and ratsnest feature. The PCB design of EAGLE can be taken as a printout in the monochromatic image form by setting a resolution of 600 dpi and can be printed on board through ultraviolet film generation or thermal printing.

12.6.4 Optimized Based System Result

The automatic light system was tested on three different input forms of light. One is in the form of an inbuilt LED of the microcontroller connected at the digital pin (D4) that blinks when given the signal. The inbuilt LED does not blink when the signal is HIGH and works opposite when the signal is LOW. Figure 12.4 describes the hardware module for remote control operation.

12.6.5 Running Android Applications

Android Studio was released at a Google conference in 2013 and made public in 2014. Earlier Android development was handled through Eclipse IDE. Android IDE is the best application to make mobile applications that end users can use on a daily basis. Android Studio helps to run the code written through an emulator or through a hardware processor (ESP8266) connected to the machine [30]. The mobile application was developed in Android Studio with login and street map activity. Functions like onCreate, onESP, onLogin, onOffline, validateEmail, Firebase authentication, etc. are created to work in modules and develop different patterns. When users open the SMART STREET application on their mobiles, firstly authorized users have to log in. A map of the university will open up and will display the poles at the location where the circuit is placed; on clicking on one of the lamps, the IP address will be displayed and then five options will be displayed as Lights On, Lights Off, ESP LED, LED, and Bulb. User can click on any one option and can see the expected result.

12.6.6 Database and Authentication Using Firebase

Light patterns are generated in the Google application called Firebase in sync with the lights used in the hardware circuit. An Android project like SMART STREET is added to Firebase by following a simple procedure by just simple login/register authentication. Firebase provides security and authentication

FIGURE 12.4
Hardware module for remote control operation.

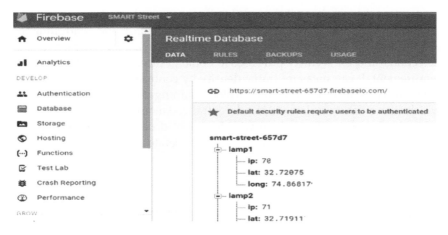

FIGURE 12.5
Authentication and database.

to user data and automatically stores user's credentials using bcrypt (a password hashing function based on Blowfish cipher) so that users can keep their focus on user interface. It supports various features like real-time database, authentication, analytics, cloud functions, etc. The application shows daily, weekly, and monthly basis of active users in graphical form and text. Any new user is added through an authentication tab; add a user by entering the mail ID of that person and password for keeping it secure. Once registered, the system generates the user ID of the particular user who has enrolled [31]. Database for the streetlight pattern is added in the database option and can be modified on a real-time basis shown in Figure 12.5.

12.7 Conclusion and Future Work

IoT is a promising technology and in trend nowadays. Strenuous efforts are put in the market to reduce human efforts and build smart environments. Smart connected devices and products provide new possibilities and opportunities for every sector, including pre-emptive maintenance to new services and business models. The case study of an automatic light system displays the remote communication of a RISC server-based processor with sensory ports and other sub-units through the Lua interface, controlled by a PDA device. The system may save up to 30% of light and energy. Authorized users are able to control the lights easily and efficiently. The system has hence reduced manual efforts and automatized the control, thus saving energy.

The above system can be developed using a solar streetlight system with automatic streetlight controller. The system gets power by harvesting the solar energy through a solar cell during the daytime with batteries and provides an effective way for the operation of the hardware circuit and controlling the light system. An automatic light system can also be used for emergency, fault detections, and for security purposes at homes, offices, or institutions as per requirement. It can measure traffic density by controlling traffic of the road. All IoT devices can be integrated with blockchaining technology like IoT chain, which allows data to be layered in a decentralized manner with improved protection and security. Blockchain with IoT-enabled devices helps in building trust between consumers and devices, and accelerates transactions.

References

1. Lee, In and Kyoochun Lee, "The Internet of Things (IoT): Applications, investments, and challenges for enterprises," *Business Horizons*, Vol. 58, pp. 431–440, 2015.
2. Roy Want, "An introduction to RFID technology," *IEEE Pervasive Computing*, Vol. 5, pp. 25–33, 2006.
3. Luigi Atzori, Antonio Lera, and Giacomo Morabito, "The Internet of Things: A survey," *Computer Networks*, Vol. 54, No. 15, pp. 2787–2805, 2010.
4. Shancang Li, Li Da Xu, and Shanshan Zhao, "The Internet of Things: A Survey," *Information Systems Frontiers*, Vol. 17, pp. 243–259, 2015.
5. Jayavardhana Gubbi, Rajkumar Buyya, Slaven Marusic, and Marimuthu Palaniswami, "Internet of Things (IoT): A vision, architectural elements, and future directions," *Future Generation Computer Systems*, Vol. 29, pp. 1645–1660, 2013.
6. H. Arasteh, V. Hosseinnezhad, V. Loia, et al., "IoT-based smart cities: A survey," *IEEE 16th International Conference on Environment and Electrical Engineering (EEEIC)*, pp. 1–6, 2016.
7. Durga Devi Sanju, A. Subramani, and Vijendar Kumar Solanki, "Smart City: IoT based prototype for parking monitoring & parking management system commanded by Mobile App," *Second International Conference on Research in Intelligent and Computing in Engineering, Annals of Computer Science and Information Systems*, Vol. 10, 2017.
8. Vijender Kumar Solanki, Muthusamy Venkatesan, and Somesh Katiyar, "Conceptual model for smart cities for irrigation and highway lamps using IoT," *International Journal of Interactive Multimedia and Artificial Intelligence*, Vol. 4, 2017.
9. Vijender Kumar Solanki, Muthusamy Venkatesan, and Somesh Katiyar, "Think home: A smart home as digital ecosystem," *Circuits and Systems*, Vol. 10, No. 7, 2016.
10. Vijender Kumar Solanki, Somesh Katiyar, Vijay Bhaskar Semwal, Poorva Dewan, Muthusamy Venkatesan, and Nilanjan Dey, "Advance automated module for smart and secure City," *Procedia Computer Science*, Vol. 78, pp. 367–374, 2016.
11. Jekishan K. Parmar and Ankit Desai, "IoT: Networking technologies and research challenges," *International Journal of Computer Applications*, Vol. 154, No. 7, pp. 0975–8887, 2016.

12. Available: https://www.arduino.cc/

13. Available: https://www.arduino.cc/en/Main/Software

14. sparkfun. What is an Arduino? Available: https://learn.sparkfun.com/tutorials/what-is-an-arduino

15. Arduino, Available: https://en.wikipedia.org/wiki/Arduino. Last modified November 29, 2018.

16. John K. Ousterhout, "Scripting: Higher level programming for the 21st century," *Computer*, Vol. 31, pp. 23–30, 1998.

17. Roberto Beauclair, "Lua Programming Language," 2011. Available: http://www.mathrice.org/IMG/pdf_Lua_Mathrice.pdf

18. Alexandre Skyrme, Noemi de La Rocque Rodriguez, and Roberto Ierusalimschy, "Exploring Lua for concurrent programming," *Journal of Universal Computer Science*, Vol. 14, pp. 3556–3572, 2008.

19. Cristina Ururahy, Noemi Rodriguez, and Roberto Ierusalimschy, "A Lua: Flexibility for parallel programming," *Computer Languages, Systems & Structures*, Vol. 28, pp. 155–180, 2002.

20. Lixia Liu, "Research on technology of embedded web server application," *2nd IEEE International Conference on Information Management and Engineering*, 2010.

21. Jyotsna A. Nanajkar and Vismita D. Nagrale, "Embedded web server based automation," *International Journal of Engineering and Innovative Technology (IJEIT)*, Vol. 2, No. 9, 2013.

22. Mohamed Abd El-Latif Mowad, Ahmed Fathy, and Ahmed Hafez, "Smart home automated control system using android application and microcontroller," *International Journal of Scientific & Engineering Research*, Vol. 5, No. 5, 2014.

23. Lalit Mohan Satapathy, Samir Kumar Bastia, and Nihar Mohanty, "Arduino based home automation using Internet of things (IoT)," *International Journal of Pure and Applied Mathematics*, Vol. 118, pp. 769–778, 2018.

24. Takeshi Yashiro, Shinsuke Kobayashi, Noboru Koshizuka, and Ken Sakamura, "An internet of things (IoT) architecture for embedded appliances," *IEEE Region 10 Humanitarian Technology Conference*, 2013.

25. Byeongkwan Kang, Sunghoi Park, Tacklim Lee, and Sehyun Park, "IoT-based monitoring system using tri-level context making model for smart home services," *IEEE International Conference on Consumer Electronics (ICCE)*, 2015.

26. Mamun Bin Ibne Reaz, Md Shabiul Islam, and Mohd S. Sulaiman, "A single clock cycle MIPS RISC processor design using VHDL," *Semiconductor Electronics ICSE*, pp. 199–203, 2002.

27. Rajeev Piyare and M. Tazil, "Bluetooth based home automation system using cell phone," *IEEE 15th International Symposium on Consumer Electronics (ISCE)*, 2011.

28. Manan Mehta, "ESP 8266: A breakthrough in wireless sensor networks and Internet of Things," *International Journal of Electronics and Communication Engineering & Technology (IJECET)*, Vol. 6, No. 5, pp. 7–11, 2015.

29. Neil Kolban, "Kolban's Book on ESP8266," 2015.

30. Android Application Development, Available: https://developer.android.com/training/basics/firstapp/index.html.

31. The Firebase website. Available: https://firebase.google.com/

13

Big Data and Machine Learning Progression with Industry Adoption

Sandhya Makkar and Vijender Kumar Solanki

CONTENTS

13.1 Introduction

There is an element of surprise when you receive a notification on your mobile that your credit card bill is due for today, or you have booked a movie ticket and it's only half an hour before the show, or there is a heavy rush on roads on the way to your office. Similarly, Facebook photo application recognizes your friends through their pictures and will send you suggestions to add them in your friend list, or your spam emails are separated from the non-spam emails automatically. Likewise, you come across various examples around you that make you wonder how your mobile phone/computer knows so much about you without you being aware about yourself. How does it send you so many alerts? The reason is your computer has learned to recognize your pattern of work and is continuously improving and learning. It is the "Google assistant" that assists you with the support of synchronise feature of 'Gmail' with it. Your computer has now learned to distinguish between the two (spam/non-spam emails) and separates them automatically. Your computer has become smart enough to recognize phone numbers and photos of your friends to suggest to you the addition of friends in your list. The technique that is working behind the scenes is "machine learning" (ML) that is supporting us. This amazing technique uses learning algorithms called neural networks, which mimic the human brain and build intelligent systems. These systems have the ability to learn from past experience or analyze historical data. They provide results according to their experience. Machine learning grew out of work in artificial intelligence; for example, it can recognize the handwriting of a person. It can prompt you to buy the things you would have scrolled and liked few days back on online shopping platforms etc. Understanding the human brain is real artificial intelligence now. In this modern era, we are making machines to think and predict. This chapter is intended to explore how the work under this technique blossomed, where is the usage of the ML in modern times, how it is effecting in revenue generation, where are its applications in industries and what are its future prospects.

This chapter is divided into five sections. Section One is discussing the introduction of ML, the section two is giving the literature work, followed by section three is about the foundation of ML followed by some famous applications in industries. Next section is on statistics of the ML, ended with conclusion and future directions.

13.1.1 Literature Review

We have identified numerous papers publications which are specifying and describing machine learning algorithms and their applications. Also, we have sought motivation to add at least one more effort toward this direction and bring a clear collection of different and unique applications-based machine learning papers. The variety of application domains, e.g., music, biology, technology, library, has been taken in account, and related to that we have covered at least one or two papers in literature to show a wide distribution work [1]. Rohan Bhardwaj, Ankita R. Nambiar, and Debojyoti Dutta have shed light on the recent outcome of how ML is working with various industries' needs and research outcomes. They have also shared the outcome on healthcare and industry initiatives to adopt the machine learning [2]. Saeed Iqbal, Muhammad Shaheen, and Fazl-e-basit presented a hypothetical and realistic examination and exploration of a number of bibliometric indicators of journal performance. A number of other factors were also taken into consideration, to prove the importance of the work. They use various ML techniques and provide a brief comparison also in the study [3]. Priyanka S. Patil and Nagaraj V. Dharwadkar expressed their idea about the banking sector and also expressed how big data is supported by machine learning. Keeping in view the importance of data as belonging to customers, they have worked on Customer Relationship Management (CRM) and fraud detection using various ML algorithms [4]. Adi L. Tarca, Vincent J. Carey, Xue-wen Chen, Roberto Romero, and Sorin Drăghici discuss machine learning with biology, how basic life science is utilizing ML, and its benefitting bioscience outcomes. Using supervised and unsupervised learning, they have given a Program Learning Outcomes (PLOs)-based study [5]. Sumit Das, Anitra Dey, Akash Pal, and Nabamita Roy have worked on ML and applications; they have shared the web page ranks and how Facebook recognizes the interests of users and shares updates through web pages [6]. Seema Sharma, Jitendra Agrawal, Shikha Agarwal, and Sanjeev Sharma have beautifully started the paper from the data mining approach and importance, and from knowledge acquisition to mapping the data is covered very effectively. They have also expressed the importance and given context to machine learning very easily [7]. Wang Hua, MA Cuiqin, and Zhou Lijuan describe the role of machine learning and also discuss the different machine learning methods starting from database populations, and then they have discussed the various methods very easily [8]. Ming Xue and Changjun Zhu have shared in their paper about the start of machine learning to various machine learning methods, for example about the rote learning, explanation-based learning, learning from instruction, learning by deduction, learning by analogy, and inductive learning.

Data is the vital chariot for machine learning algorithms. Many researchers contend that despite promises of big data's potential value to organizations, it remains an untapped resource in many business sectors [9]. Although new technologies have seen an increase in the supply of data, many organizations in different industries still struggle to reap the benefits of big data [10]. The generation of huge data from different sources such as tablets, smartphones, sensors, and the Internet has led to an overwhelming growth of unstructured data difficult to process with traditional technologies [11]. Companies are facing issues of handling the huge amount of data [12–14]. So, this has become a big challenge to them to amalgamate and sort the data for their processes. It is very important for organizations to manage big data effectively to get the benefits [15], which are not always obvious. Soanki et al. provided IoT based predictive models for car maintenance, smart cities etc. [16–19]. Although many researchers have shown a correlation between effective use of big data and financial performance, evidence of big data's value to many organizations remains obscure [20].

In the following sections of the study the historical view of the AI techniques, industry paradigms and their revenue generations are discussed. In Section 13.4, machine learning development and adoption life cycle is presented, following with the section explaining the bottlenecks and challenges faced by firms during the adoption. And, finally, the last section focuses on future directions and concluding remarks.

13.1.2 Historical Sketch of Machine Learning

In 1946 the first computer system was developed. At that time the word "computer" meant a human being who performed numerical computations on paper. This machine was manually operated; i.e., a human would make connections between parts of the machine to perform computations. The idea at that time was that human thinking and learning could be rendered logically in such a machine. Some of the remarkable events in machines as collected by Forbes [21] are mentioned in the following and is vital to mention here to fully understand the story of big data and machine learning (Table 13.1):

TABLE 13.1

Remarkable Events in History of Machine Learning

Years	Description
1950	Alan Turing creates the "Turing Test" to determine if a computer has real intelligence. To pass the test, a computer must be able to fool a human into believing it is also human.
1952	Arthur Samuel wrote the first computer learning program. The program was the game of checkers, and the IBM computer improved at the game the more it played, studying which moves made up winning strategies and incorporating those moves into its program.
1957	Frank Rosenblatt designed the first neural network for computers (the perceptron), which simulates the thought processes of the human brain.
1967	The "nearest neighbor" algorithm was written, allowing computers to begin using very basic pattern recognition. This could be used to map a route for traveling salesmen, starting at a random city but ensuring they visit all cities during a short tour.
1979	Students at Stanford University invent the "Stanford Cart" which can navigate obstacles in a room on its own.
1981	Gerald Dejong introduces the concept of Explanation Based Learning (EBL), in which a computer analyzes training data and creates a general rule it can follow by discarding unimportant data.
1985	Terry Sejnowski invents NetTalk, which learns to pronounce words the same way a baby does.
1990	Work on machine learning shifts from a knowledge-driven approach to a data-driven approach. Scientists begin creating programs for computers to analyze large amounts of data and draw conclusions—or "learn"—from the results.
1997	IBM's Deep Blue beats the world champion at chess.
2006	Geoffrey Hinton coins the term "deep learning" to explain new algorithms that let computers "see" and distinguish objects and text in images and videos.
2010	The Kinect can track 20 human features at a rate of 30 times per second, allowing people to interact with the computer via movements and gestures.
2011	IBM's Watson beats its human competitors at Jeopardy. Google Brain is developed, and its deep neural network can learn to discover and categorize objects much the way a cat does.
2012	Google's X Lab develops a machine learning algorithm that is able to autonomously browse YouTube videos to identify the videos that contain cats.
2014	Facebook develops DeepFace, a software algorithm that is able to recognize or verify individuals on photos to the same level as humans can.
2015	Amazon launches its own machine learning platform. Microsoft creates the Distributed Machine Learning Toolkit, which enables the efficient distribution of machine learning problems across multiple computers. Over 3,000 AI and robotics researchers, endorsed by Stephen Hawking and Steve Wozniak (among many others), sign an open letter warning of the danger of autonomous weapons which select and engage targets without human intervention.
2016	Google's artificial intelligence algorithm beats a professional player at the Chinese board game Go, which is considered the world's most complex board game and is many times harder than chess. The AlphaGo algorithm developed by Google DeepMind managed to win five games out of five in the Go competition.

2016 may very well go down in history as the year of "the Machine Learning hype." Google announced they were open-sourcing Tensor Flow (TF)! TF is already a very active project that is being used for anything ranging from drug discovery to generating music. Microsoft open-sourced CNTK, Baidu announced the release of Paddle Paddle, Google is also supporting the highly successful Keras, so things are at least even between Facebook and Google on that front. Some of the popular industry paradigms are discussed in the following section.

13.2 Machine Learning and Industry Paradigms

According to Gartner, machine learning is #1 on the list of the top 10 strategic technology trends for 2017, closely followed by intelligent apps and intelligent things. This just goes to show us these devices aren't going away any time soon.

Firms always work for revenue generation and the applications of machine learning in various areas brought growth in their various verticals. The revenue generation is discussed in the following section.

Company	Technique Adopted	Description
Google (Alphabet)	Visual Object Recognition Virtual Personal Assistant GPS Navigation Services Email Spam and malware filtering using Decision Tree Induction of ML Chatbot Search Engine Result Refining	Face recognition and character recognition. Assist and instruct in finding information integrated with smart speakers, smart phones and mobile apps. Traffic predictions, Spam filters track the latest tricks adopted by spammers and filter the emails. To extract information from website and present it to the customer. Estimate the search results according to the requirement.
Amazon	Product Recommendation	Prompts and emails about the products liked by the prospective customer.
GlaxoSmithKline—a global healthcare company	Luminoso's Natural Language, Text Analytics Technology	To tease out parent's fear that links vaccination and disease to autism.
ZenDesk	MarianaIQ's Social Media Engagement Platform	Identify the patterns in contact data and target those audiences that are ready to purchase their products.
Siri	Virtual Personal Assistant	Assist and instruct in finding information integrated with smart speakers, smart phones, and mobile apps.

(Continued)

Company	Technique Adopted	Description
Facebook	People You May Know Face Recognition Chatbot Army	Friends suggestions on basis of your visiting their profiles, similar workplace, interests, etc. Face detection on the picture and prompting to tag. Chatbots are included in Facebook (FB) Messenger and any developer can create and submit a chatbot for inclusion in FB Messenger to serve their customers better.
twitter	Curated Timelines	To show "the best tweets first." Scoring tweets according to various metrics based on individual preferences.
Uber	Online Transportation Networks	Price estimation, minimizing the detours.
Pinterest	Similar Pins/Improved Content Discovery	Identifies, extracts and recommends the useful information from images.
Paypal	Cyberspace Fraud Detection	Protection against money laundering to detect illegitimate transactions between buyer and seller.
Baidu, the Chinese search engine	Speech recognition/translation using Deep Voice Technology	Generates synthetic human voices that are very difficult to distinguish from genuine human speech.
IBM	IBM Watson	Watson has been deployed in several hospitals to detect certain type of cancers. Watson is used as an assistant to help retail shoppers.
Salesforce	Salesforce Einstein	Analyze every aspect of customer's relationship for giving effective customer service.
Edgecase	Improving Ecommerce Rate	Providing better shopping experience to retailers on online platform. Streamlined and improved conversion rates.

13.3 Industry Revenue Generations

This section discusses the revenue generation by the firms using machine learning algorithms and adopting big data in their processes. It is logical that the number of customers are directly proportional to the revenue generations; i.e., more customers, more profit. But the intriguing question that concerns managers of today is how installing and implementing artificial intelligence, big data and machine learning technologies will impact the overall revenue generation processes of the organization. In this section, we have attempted to provide details of the same.

Machine learning can bring out improvement in revenue generation, though not quickly but slowly and steadily. Some of the reasons for increase in revenue generation are (a) ML technology can facilitate organizations to tap new customers by anticipating market trends, new sales methodologies, and quick response to the customers; (b) the use of chatbots and algorithms to dig deep into customers' likes and dislikes can improve customer satisfaction and can further result in steep growth of revenue; (c) ML helps in financial modeling through which a customer's expenses can be tracked, better sales forecasts can be predicted and chances of errors can be minimized; and (d) with proper monitoring and scrutiny using ML and computing algorithms, operational costs of business systems can be lowered. So, ML gives the organizations a platform for growth in revenue generation.

CEOs and CIOs are now aware of the fact that more data means more value to the company. According to the research by the Economist Intelligence Unit, 60% of the professionals feel that data is generating revenue within their organizations and 83% say it is making existing services and products more profitable. According to a report by Forbes [22]. Big Data, adoption reached 53% in 2017 up from 17% in 2015, with telecom and financial services leading early adopters.

According to report by Forbes [17)] machine learning patents grew at a 34% Compound Annual Growth Rate (CAGR) between 2013 and 2017. International Data Corporation (IDC) forecasts that spending on AI and machine learning will grow from $12B in 2017 to $57.6B by 2021. Deolloitte Global has predicted the number of machine learning pilots and implementations will be doubled in 2018 compared to 2017 and will further double itself by 2020. In the coming years, there will surely be a leap in the usage of ML algorithms, and huge numbers of corporations will be using this technology to upscale their business.

Despite of number of promising points for revenue generation by ML, it is also necessary to understand what it is that is holding ML back for application purposes in industries. According to the CIOs and CEOs there are far less real practices in the industry, so an implementation is difficult without a real human mind. Other reason is the ML tools and technology are too young and they are still evolving, so they may not fit with the complexities of the businesses till now. A very valid reason is that it's very difficult to collect data to work on. Also, the models that are used to predict and forecast are not easy to understand and require specialized training. And not the least, there are major privacy concerns with the advent of the new technologies.

13.4 ML Development and Adoption Life Cycle

It has now become essential for the organizations to reengineer their old processes that required them an in-depth discovery and exploration on a single project, and later, validating, checking the feasibility, debugging, and finally implementing it after a long wait of months. Embedding ML-based models make products quicker, more economical, increase ROI, and personalize them according to the needs of twenty-first-century smart customers. For the adoption of the emerged techniques of machine learning, industries need to map its processes with the phases of ML techniques and must understand the challenges and bottlenecks they may face during the deployment. Before moving ahead into the life cycle of adoption of ML, it's important for organizations to ensure a pre-study of the internal environment of the organization and its characteristics. According to [23] organizations must identify the following beforehand: the need and importance of adopting ML techniques, satisfaction with the existing system, familiarity with ML techniques and competencies in ML applications. In this section, we are presenting a general framework for the adoption of ML techniques into the processes. Organizations must understand the need and importance of adopting. The other factors that organizations must study are the external ones; i.e., what their competitors are doing and what is happening in the market or in other industries. Last but not the least, firms must study the characteristics of ML itself, as adoption of new technology is heavily dependent on the characteristics of the new tools/technologies. Organizations must consider the perceived benefits, perceived barriers and tools' availability. The snapshot of the above description can be seen in Figure 13.1.

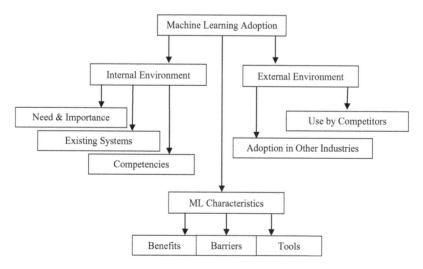

FIGURE 13.1
Before adoption: Set-up in industries.

Since machine learning is a method of data analysis that automates analytical model building based on the idea that systems can learn from data, identify patterns and make decisions with minimal human intervention, therefore, the availability of massive amounts of data becomes the key driver in adoption of ML by industries. This becomes double-fold by the growing demand for superior customer service, efficient operations and thereby improved sales revenue.

Machine learning is the process of extracting data, feeding into machine learning engines, and then generating models. These models are then injected in business processes to generate the results. One can easily interpret the life cycle by understanding Figure 13.2.

After data extraction from various in-house sources or other external sources (Relational Database Management System (RDBMS), Simple Storage Services (S3), Hadoop Distributed File System (HDFS), etc.) that require data engineering, the processes become data science–heavy. Figure 13.3 shows the two phases of ML adoption, "Building ML" and "Applying ML," in detail. Both phases require data science processes as well as expertise in software engineering. Data scientists work on data, predict the patterns using various ML engines like penalized linear regression, decision trees, and neural networks. The main focus of data scientists is to use and apply as many machine learning engines as needed along with various algorithms to solve a specific problem. Figure 13.3 gives the conceptual framework of various engines. The software engineers' work is also critical in ML processes (both building and applying). They support the team of data scientists in speeding up and refining the processes and make the recurrent work easy in building the model.

Since the development stage is data science–heavy, we can approximate the work of data scientists versus that of soft engineers in a ratio of 80:20, and other stages of applying and speeding up in a ratio of 30:70. The outcome of the "Building ML" stage can be directly injected into the business processes

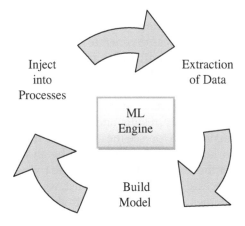

FIGURE 13.2
Lifecycle of ML.

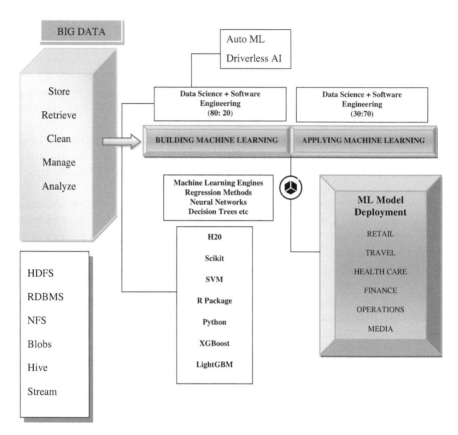

FIGURE 13.3
ML building and application phase.

and thereby can be used by consumers, enterprises as a final product for business intelligence, and revenue generation. Or else, the product is injected into another ML building process to generate further results. These processes are general stages that any organization can follow to build and adopt ML into their function of business [24].

The promise of machine learning is real and so are the challenges. With the race in adoption of machine learning computing due to its rising importance because of speedy results and optimized processes at a lesser cost, organizations may fail to foresee the challenges that they may face both before adoption and later on. This section will be an eye-opener for those facing these issues for the first time and also that are trying to embed machine learning into their processes. In the following list, we have tried to group the challenges that companies are facing along with the probable solutions.

1. **Inaccessible data:** ML is the system that is fueled by the data. Therefore, data set is a vital component for an epic machine learning

implementation. Many times, it becomes difficult for the firms to acquire the data or access the right amount of data that is required to predict the patterns. This could be the biggest challenge for any of the firms. A data acquisition can be done from registered firms authorized to sell the relevant data. Data can be collected from Enterprise Resource Planning (ERP) databases, mainframes or instrumented devices that are part of IoT (Internet of Things) systems.

2. **Data security:** The next and the vital issue can be the security concerns. Once data acquisition bottleneck is resolved, one must be cautious about the security of this data. Firms should be able to identify which data is of prime importance and which is not. The sensitive data must be accessible only to a few involved in the ML adoption.

3. **Infrastructure requirements:** Machine learning is still in the budding stages in many developing countries. It requires the usage of testing and experimenting time and time again to get the best possible results. This requires the hiring of those with relevant skill sets or the resources for best deployment. Apart from that there should be a big investment in IT softwares to support ML deployment. Machine learning feeds off of large amounts of raw data, which often come from various sources. This brings a demand for advanced data integration tools and infrastructure, which must be addressed in a thorough data integration strategy.

4. **Inflexible business models:** Implementations may not be successful every time. For the successful implementation, agile and flexible business processes are crucial. The companies also need to spend less time, effort, and money on unsuccessful projects. If one of the ML strategies don't work, it enables the company to learn what is required and consequently guides them in building a new and robust ML design.

5. **Affordability for organizations:** According to O'Reilly, it was found that the average base salary of a data scientist in the United States in the year 2014 was $105,000. This figure does not include benefits, bonuses, or any other compensation. Including all the other expenses, the figure can rise up to $144,000. There is a virtue in knowing these values if you're looking to implement machine learning, because if you're applying machine learning, you will require data engineers and a project manager with a sound technical background. In essence, a full data science team isn't something newer companies or start-ups can afford.

6. **Complexity in ML algorithms:** According to a study of Gartner, [25] organizations often spend too much time debugging models that don't fit the data, business problem or challenge they are trying to address. Start by categorizing to help reduce capabilities and to avoid overwhelming users.

Apart from above challenges, the deployment and adoption stages can have challenges like of problem identification and right formulation. A wrong problem formulation would lead to dead ends: performing data engineering, cleaning, feature extraction/selection etc., applying the right ML engines, and choosing proper parameters to tune algorithms lead to performing fair experiments. It is suggested that firms must invest in a small ML platform first, based on and designed to support your initial use, and iteratively expand the ML platforms over time. Firms must figure out the basis to improve data quality; for this, digitization of the processes is a must. It is also important to know what areas need automation support. Also, will there be any gain in productivity if those processes would be automated using ML algorithms? Also, boosting the operational efficiency should not only result in revenue generation: one must build an exceptional customer experience. Hiring and training before the implementation is necessary. And the goal should be clear to each one in the organization and should not be restricted only to CIOs.

13.5 Future Directions and Concluding Remarks

In the next ten years, there will be a few areas where there would be a huge impact of machine learning all over the world. Netflix, Google assistant, Facebook, etc. alerts are just the beginning of machine learning. There will be better decision-making processes. There will be a revolutionary personalization, from services to the products. Consider the creation of specialized products that are customized specifically for you. With the tons of data and data analytics, it will bring you the shampoo you would like to use and that suits your hair. If we talk about education industry, currently the computer plays a role of a server for finding, moving, translating, calculating the numbers, and you have to memorize the rest of the things. But with the use of machine learning, the focus on memorization and writing would not last. In the blink of an eye, the information you required would be in your chip fitted in your eyes like a contact lens and creation and innovation would be the motto using the information.

There would be massive cultural transformation in the coming decade. Focus will move from resistance to the use of computers to the expectation that machines to liberate and allow us to design, create, and innovate. Machines will be the partners in the processes. It will be the cultural revolution of the next decade. While many of the stories in the past couple of years have been about how machine learning can be used to analyze images and video, the applications relating to human voice are at least as important. The new iPhone may be able to unlock itself when it recognizes your face, but what you'd probably find more useful is if Siri could save you from all the typing you have to do—for work and in your personal life. It does seem an extremely tricky challenge, though, since not only is each of our faces unique; our voices and the way we speak are also particular to us. But AI and machine learning have many other applications beyond analyzing images and sounds, so it will be interesting to see how the world is changed by it.

According to one of the studies of Forbes [26], enterprise investments in machine learning will nearly double over the next three years, reaching 64% adoption by 2020. IDC is forecasting spending on artificial intelligence and machine learning will grow from $8B in 2016 to $47B by 2020. Eighty-nine percent of CIOs are either planning to use or are using machine learning in their organizations today. Fifty-three percent of CIOs say machine learning is one of their core priorities as their role expands from traditional IT operations management to business strategists. CIOs are struggling to find the skills they need to build their machine learning models today, especially in financial services.

The usage of ML is not limited to these few applications. The growth of ML would be unlimited, and there would not be any industry vertical that would be beyond the scope of ML, be it healthcare, entertainment, financial services, manufacturing, airlines, and more. Companies are revamping the processes and skills of their employees. Over half (56%) of responding companies use third parties to develop machine-learning capabilities, while 17% rely on their regular IT process. So, there is promising growth of overall industry with the inception of machine learning ten years down the line. By the year 2030, ML will have changed the way we travel to work and to parties, how we take care of our health and how our children would be educated.

References

1. R. Bhardwaj, A. R. Nambiar, & D. Dutta, "A Study of Machine Learning in Healthcare," *2017 IEEE 41st Annual Computer Software and Applications Conference*. 2017.
2. S. Iqbal & M. Shaheen, "A Machine Learning Based Method for Optimal Journal Classification," *The 8th International Conference for Internet Technology and Secured Transactions (ICITST-2013)*. 2013 (IEEE), London UK.
3. P. S. Patil & N. V. Dharwadkar, "Analysis of Banking Data Using Machine Learning," *International Conference on I-SMAC (IoT in Social, Mobile, Analytics and Cloud) (I-SMAC 2017)*.
4. A. L. Tarca, V. J. Carey, X.-W. Chen, R. Romero, & S. Dra"ghici, "Machine Learning and Its Applications to Biology," *PLOS, IEEE*. Open Access.
5. S. Das, A. Dey, A. Pal, & N. Roy, "Applications of Artificial Intelligence in Machine Learning; Review & Prospect," *IJCA* (0975–8887), 115(9), 2015, pp. 31–41.
6. S. Sharma, J. Agrawal, S. Agarwal, & S. Sharma, *"Machine Learning Techniques for Data Mining: A Survey," School of Information Technology*, UTD, RGPV, Bhopal, MP, India, 2013.

7. H. Wang, C. Ma, & L. Zhou, "A Brief Review of Machine Learning and Its Application," 2009 International Conference on Information Engineering and computer 2009.

8. M. Xue & C. Zhu, "A Study and Application on Machine Learning of Artificial Intelligence, *2009 International Joint Conferences on Artificial Intelligence.* 2009.

9. Beyer, M.A. and Laney, D. (2012) The importance of "Big Data": A definition Gartner, G00235055.

10. K. Crawford, "The Hidden Biases in Big Data," *Harvard Business Review,* 2013. Show Context. https://hbr.org/2013/04/the-hidden-biases-in-big-data.

11. D. Bollier, "The Promise and Peril of Big Data," Aspen Institute, 2010. Show Context. A report by aspen institute, washington, DC, USA.

12. https://www.marutitech.com/challenges-machine-learning/.

13. E. Carlton, Sapp. https://www.gartner.com. preparing_and_architecting_for_machine_learning.pdf, last retrieved on July 5, 2018.

14. https://www.forbes.com/sites/louiscolumbus/2018/02/18/roundup-of-machine-learning-forecasts-and-market-estimates-2018/#51a044822225.

15. E. Brynjolfsson, L. M. Hitt, & H. H. Kim, "Strength in numbers: How Does Data-Driven Decision Making Affect Firm Performance? ebusiness.mit.edu," 2011. https://ssrn.com/abstract=1819486.

16. R. Dhall & V. Solanki, "An IoT Based Predictive Connected Car Maintenance," *Approach International Journal of Interactive Multimedia and Artificial Intelligence (ISSN 1989–1660).* Vol 4, No- 3, 2017, pp. 16–22.

17. V. K. Solanki, M. Venkatesan, & S. Katiyar, "Conceptual Model for Smart Cities for Irrigation and Highway Lamps using IoT," *International Journal of Interactive Multimedia and Artificial Intelligence, (ISSN 1989–1660).* Vol 4, No- 3, 2017, pp. 28–33.

18. V. K. Solanki, V. Muthusamy & S. Katiyar, "Think Home: A Smart Home as Digital Ecosystem," *Circuits and Systems,* 10(07), pp. 1976–1991, 2016.

19. V. K. Solanki, S. K., V. B. Semwal, P. Dewan, M. Venkatesan, N. Dey, *Advance Automated module for smart and secure City in ICISP-15,* organised by G. H. Raisoni College of Engineering & Information Technology, Nagpur, Maharashtra, on December 11–12, 2015, published by Procedia Computer Science, Elsevier, Vol 78, pp. 367–374.

20. D. D. Sanju, A. Subramani, & V. K. Solanki, "Smart City: IoT Based Prototype For Parking Monitoring & Parking Management System Commanded By Mobile App," *In Second International Conference on Research in Intelligent and Computing in Engineering.*

21. B. Marr, "A Short History of Machine Learning Every Manager Should Read," *Forbes.* https://www. forbes. com/sites/bernardmarr/2016/02/19/a-shorthistory-of-machine-learning-every-managershould-read. Retrieved September 28, 2016.

22. https://www.forbes.com/sites/louiscolumbus/2017/10/23/machine-learnings-greatest-potential-is-driving-revenue-in-the-enterprise/#4c1e604841db.

23. https://ieeexplore.ieee.org/stamp/stamp.jsp?tp=&arnumber=7293887&tag=1.

24. A. Chauhan, "How to Adopt Machine Learning," https://www.dzone.com, last retrieved on July 4, 2018.

25. https://www.gartner.com/doc/3471559 last accessed on April 30, 2018.

26. https://www.forbes.com/sites/louiscolumbus/2017/10/23/machine-learnings-greatest-potential-is-driving-revenue-in-the-enterprise/#614da2b041db last accessed on April 30, 2018.

14

Internet of Things: Inception and Its Expanding Horizon

Swagat Rameshchandra Barot, Subhanshu Goyal, and Anil Kumar

CONTENTS

14.1 Introduction

"Internet of Things" (IoT) is popularly defined as the grid of tangible devices including automobiles, household instruments, and other things equipped with software, electronics, sensing devices, and Internet connectivity, which empower these items to communicate and transfer information, giving rise to possibilities for more direct amalgamation of the real world into computing devices, leading to betterment in efficiency, financial gain, and less human effort. It is also defined as a *network of various computing devises connected to internet and implanted in objects of routine usage to empower them to exchange information.* IoT is an abbreviation of Internet of things which is related to an ever-growing network of tangible objects possessing IP address for connectivity through internet, and the exchange of information taking place between these tangible objects and other items and/or systems capable of connecting through internet. In non-technical laymen's terms, any network of physical things with Internet and computing devices that produce, transfer, and occasionally even analyze the information leading to the ease of human life and jobs can be considered as an Internet of Things.

14.2 Historical View of IoT

Even though the buzz of IoT has been started showing up for quite a few years and we have been seeing this concept being taught in universities, the concept of the IoT is not that novel. The first noteworthy public discussion about the concept was heard from Nikola Tesla in 1926, when in an interview he very confidently predicted that in the very short future the concept of "wireless" would be widely and wisely applied, the that the earth would be an imitation of human brain on a giant basis. He justified the metaphor of the human brain by showing the similarity that everything on the earth was element of physical and periodic whole. He went further to predict that the machinery components of the mechanism would be extremely simple. Elaborating the simplicity of the machinery components, he said that it would be no more complex than the telephone of that time. He further foresaw that such mechanisms would be so common and tiny that almost every individual would have them in their trouser pockets. Similar thoughts were expressed by Alan Turing, the father of modern computing, in 1950, through one of his articles entitled "Computing Machinery and Intelligence" [1]. He forecasted that computing machines could be trained to imitate the human brain, provided they are well equipped with the best possible sensory mechanism and could be taught to understand and speak language. He found it very much possible that computing devices could be trained to learn new things like human children. In 1982, the world saw a first example of a smart machine when the research team of the computer science department of Carnegie Mellon University mounted micro-switches in the vending machine to keep a watch on the number of Coke bottles left unsold in the vending machine and the temperature of those bottles. In 1989, a project manager,

Chriet Titulaer, of the Netherlands attracted the attention of the entire world by presenting an example of a smart home, which had given rise to many technological inventions to enable effective interactions between human and home appliances. In fact, the concept of voice recognition was also a striking feature of the home. The beginning of connecting machines to the Internet nonetheless began in 1990, when Simon Hacket and John Romkey developed a toaster connected to the Internet (which they called Sunbeam Deluxe Automatic Radiant Control Toaster), which couldn't do much except get switched on and off automatically. The improved version of the toaster came in the next year, wherein the toaster could automatically take in the bread slice for toasting.

However, according to many references, official research pertaining to something very close to what is nowadays known as Internet of Things (IoT) started in 1999 [2]. A research team of a United Kingdom's technocrat Kelvin Ashton (who was later given the title of *the father of IoT*) comprised of two MIT professors, Sunny (Kai-Yeung) Siu and Indian origin professor Sanjay Sarma, started working in a laboratory of MIT. The team was researching to find ways to connect the Radio Frequency Identification (RFID) to the World Wide Web in the AutoID laboratory. Mainly involved in the research problem of the field of logistics and supply chain management, the team searched for the solution of the problem from technology, which ultimately gave rise to the field of IoT. Actually, Kelvin Ashton, an assistant brand manager in the multinational company Proctor & Gamble, once found that the inventory management handled by human intervention aided technology was not doing well, which led him to the research project conducted by the team [2]. Kelvin Ashton could see that the solution might come from the understanding of a similar problem in the nature; that is, the human sensory system. Working on the same theme, they could conclude that unlike the human sensory system, where the brain sees the world with the data provided by sensory organs, an intelligent system needs to get the capability of sensing in order to imitate the human sensory system. In 1999 speech, Kelvin explained how a majority of the data (worth 1024 terabytes) available on the entire World Wide Web by that time was created and posted on the Internet by humans, which required enumerable human hours [2]. If computers or computing systems were empowered to imitate humans and to know everything about the world on their own, without humans entering the information about it in any way, the wastage of human time could be saved. At the same time, information would be more accurate and cheaper. The nomenclature of the concept was stirred by the name of the book *When Things Start to Think*, written by Dr. Gershenfeld, a professor at the Massachusetts Institute of Technology. In his first presentation about the concept, Kelvin entitled his presentation as "Internet of Things (IoT)," which later on became the name of the concept [2].

14.3 Nontechnical Definition of Internet of Things (IoT)

In simple terms, when a number of daily life things are connected among one another via the Internet so that it can generate, transfer, and analyze data and use it for easing human life, they can be called smart machines or the Internet of Things. To be more technical, there are three main components of any IoT infrastructure: sensors and actuators, linkage, and people and process. We may find a number of examples of IoT around us:

- Mobile applications related to health and fitness are an embodiment of the Internet of Things. The applications use the Internet and GPS, monitor your activities using them, calculate the numbers of steps traveled, total running/walking time during the day, total calories burned during the day, and so on. It keeps track of all these data and accordingly gives dietary suggestions too.

- Smart air conditioners are also one type of IoT device. Based on the temperature outside the room, the season and many other factors, it modifies and maintains the temperature of the room. In fact, it also directs the swing fan in the direction of the individual present in the room.

- Smart lighting is a system that regulates the color and intensity of the light and can be controlled through mobile phones.

- A smart transportation system is a network of vehicles in which vehicles are GPS-enabled and interconnected so that locations of vehicles can be known by the other vehicles too.

- Smart lock is an intelligent locking system, which is a lock with sensors and doesn't require a key to open and lock it. In fact, the system has multiple features, for example, keeping track of entries and exits, allowing listed people to have free/restricted access, remote operation of the system through mobile phones, and more. The system can also provide reminders to the owners about several outdoor tasks.

- As mentioned earlier, the concept of the smart home turned into a reality long ago. However, due to continuous improvement and amendment in the mechanism, the recent smart homes include multiple features ranging from auto-cleaning to auto-cooking mechanisms. Self-regulatory safety mechanisms (similar to smart locks) and anti-fire systems are also features of smart homes of the current era.

14.3.1 Beneficiaries of Internet of Things

There are number of sectors that get benefited by IoT. In fact, one can say that there is no sector associated with human life that has remained untouched by IoT.

- **Health sector:** For regular monitoring of health, remote diagnosis and surgery, and dietary monitoring, IoT can play a vital role. In case of psychological and disabled patients, IoT can provide regular monitoring of health parameters, namely, blood sugar, blood pressure, and other health-related indices.

- **Transportation system:** IoT can play a crucial role in complete automation of automobiles, which can cut down the chances of accidents due to human error. Going one step beyond, IoT can change the entire transportation mechanism. Integration of GPS to automobiles can provide directions to vehicles about which is the shortest and most traffic-efficient route, seeing the real-time traffic situation. It may also be improved to maintain traffic situations and provide traffic instructions to each vehicle directly, which would eradicate the requirement of traffic police and traffic signals at all.

- **Manufacturing and production:** Several manufacturing and production operations like process control, regulation of safety parameters, and overall monitoring of the production unit could be automated by IoT infrastructure.

- **Agriculture and irrigation:** Through effective usage of the IoT, each plant can be provided individual care and treatment based on its requirement, depending on different parameters related to it; namely, amount of fertilizer and water required and the time of providing it, weeding, immediate pest control, and more. Accurate data related to plant can be furnished and processed through sensors connected to plants.

- **Optimization of energy consumption:** The majority of energy wastage can be regulated using the IoT, as it can sense the wastage at the initial stage and can take corrective actions automatically. As each component of the system sends data to the IoT, energy wastage at any juncture could be identified and regulated by the IoT.

14.4 Major Gains Due to IoT

- **Easy interchange of information among machines:** Due to the IoT, a variety of machines/devices can communicate with each other and can work more cohesively. For example, in case of accidents, a road CCTV camera and associated computer vision unit can immediately send a message to the ambulance and nearby medical team to be ready for appropriate actions. The visuals and current status of the patient could be constantly communicated to the team of doctors during the shifting of the patient to the hospital, so that they can immediately give appropriate treatment as soon as the patient arrives at the hospital. Automated wheelchairs and automatic inventory management of different medicines and oxygen cylinders resulting from intercommunications among devices can completely change the way hospitals and doctors work.

The devices can also be controlled from remote places: the IoT can enable devices to be effectively controlled from remote places. For example, an engineer can remotely give directions to the production unit in case of need. Even a smart locking system of the home can be regulated from the remote location by its owner.

- **Reduction in cost:** Due to automation, the IoT will reduce the chance of human error and will reduce the chance of loss of many kinds, be it human life or money. In production units, timely action in case of failure of the unit will quickly resolve the problem and can resume the production in no time, which may save the factory from potential financial loss. Due to smart transportation systems, automobiles will take short and traffic-free routes, which will save the fuel and hence the cost.

14.5 Rapid Growth of Internet of Things

The data shows that the notion of the Internet of Things is being adopted and applied to various fields worldwide. Graph 14.1 shows the rapid growth of it at global level by means of amount of dollars spent on it.

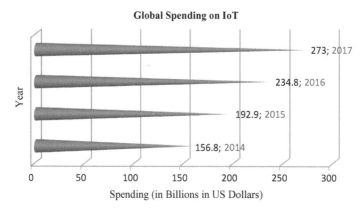

GRAPH 14.1
Global spending on the Internet of Things.

According to General Electric (GE), in the world's overall GDP, approximately 10^{13} to 1.5×10^{13} US dollars will be contributed by the ventures related to IoT by 2030. According to [3], global spending on the IoT is growing at the rate of 16.7% per year and has crossed 6×10^{13} in the year 2017. The growth is expected to boom in further years as private and public establishments and governments are spending in the IT sector, services, and connectivity enabling IoT.

According to the same survey, about 1.4×10^{12} US dollars are expected to get spent globally by 2021. Among these, with the spending of 4.55×10^{11} US dollars, the Asian Pacific region is likely to top the tally, leaving the US behind with the likely spending of 4.21×10^{11} US dollars. Western Europe is likely to spend 2.74×10^{11} US dollars.

In 2016, in the top IoT firms, spending on manufacturing and operations was 1.05×10^{11} US dollars out of the total spending of 1.75×10^{11} US dollars. Shockingly, the spending on manufacturing and operations was higher than that of production, asset management and field service.

14.6 Possible Future Applications and Directions

As discussed previously, each and every facet of human life has the potential of getting changed by the IoT, and our communication capability will not remain limited to only mobile devices. Rather, it will spread to each and every thing we deal with. The Internet of Things can affect each facet of human life by its applications. A number of applications are mentioned in the literature [4]. A few of these applications are as follows:

1. **Forecasting of natural calamities:** The blending of sensors and their independent coordination and simulation will help to give warnings against any natural calamities and to take proactive moves in advance.

2. **Applications to industry:** There are a number of industrial applications of the IoT; for example, management of all automobiles in the parking lots of an organization. The IoT is useful in evaluation of environmental performance of the industry and is also useful for data processing to identify those which need maintenance.

3. **Monitoring on water shortage:** The IoT may be useful in identifying water shortages at various places. A collection of sensors embedded in the network, along with the specific simulation activities, might not only keep an eye on long-term water interventions but also keep a close watch on accidental mixing of sewage water with the regular water stream/storage, which might otherwise lead to harmful results on the health of humans.

4. **Smart homes:** The concept of smart homes can be actualized by means of the IoT. The defining features of smart homes include optimization of energy consumption, intercommunication of home amenities, home safety features and automatic actions in case of emergencies [5,6].

5. **Medical applications:** The Internet of Things can play a vital role in the field of medicine, for example, determining various health-related measurements like BP, sugar levels, thyroid condition, monitoring bodily activities, support for independent living, regular alerts and help in medication consumption.

6. **Agriculture application:** Through a network of sensors installed in farms, data can be produced, processed, and analyzed and can yield inferences related to farming, which can be sent to the farmers; for example, the farmer can be alerted by the IoT infrastructure in case of the need for fertilizer in a particular region of a farm. Alerts for providing water to the crop can be provided to the farmer through moisture sensors installed in the drip irrigation system. IoT infrastructure can provide alerts for pest control too.

7. **Design of smart transportation:** The IoT can contribute to the design of effective and smart transportation systems [4]. By harnessing the power of the Internet, GPS, and sensors network, the smart transport system can give solutions to many existing problems of transportation, which includes traffic jams, road safety, traffic violations, pollution, and time accuracy and time management of public transport. The smart transportation system may guide vehicles to choose the least traffic-bearing routes to reach destinations in less time and in this way can help the spot with the traffic jam by getting rid of additional vehicles. In fact, the IoT-enabled transportation system can immediately inform the concerned authorities about the traffic jam so that corrective actions can be taken in minimum time. The system may solve the problem of traffic jams at the toll plaza by the integration of electronic toll payment systems in the vehicles. The system could closely monitor violation of traffic rules by CCTV cameras and could issue electronic traffic tickets too. By immediately informing an ambulance in case of an accident, the system can save precious human lives too [7].

8. **Embodiment of the concept of smart city:** The concept of a smart city can be actually implemented by taking the help of IoT infrastructure. Such smart cities will include smart transportation system, automatic streetlight regulation systems to save energy, energy-saving systems, smart communication systems, smart waste management systems, e-payment systems for various tasks, and so on [3,8]. The striking features of the smart cities are easing human lives by effective usage of technology that could be architected by the IoT.

9. **Smart billing system for utilities:** By the use of the IoT, utility distribution and billing can work intelligently. Such a smart system can

easily catch any power theft in the electricity distribution system by measuring the electric flow at various levels through sensors. The billing can also be automated, so that the accurate meter reading can take place and the bill can be issued through e-mails and SMSs at each month's end. Users can be provided online payment facilities, which can put a heavy cut on the cost borne by such organizations on physical bill payment centers. Such system can also bring reduction in wastage of water, electricity and gas during the distribution by preventing and effectively dealing with leakage in the distribution system.

10. **Security management system:** The IoT has a huge scope of contributing to the field of home security systems and surveillance of various public and private organizations. The IoT can provide features like identity verification system, regulated entry-exit system, alarm in case of unauthorized entry, and so on, which can strengthen the security of the organization.

In addition to the above-mentioned applications, the IoT has gained the attention of academia, industries, and governments at large. The world's top universities, a majority of which are funded by industries and governments, are involved in the research related to IoT. Many international organizations are engaged in specific IoT development. Many corporate giants are helping governments and societies at large by providing IoT infrastructures to effectively address water and air pollution problems. To name one, the Eye-on-Earth platform by Microsoft is focusing on water and air quality problems of many European countries. This program significantly helps ongoing worldwide research in the field of environment studies by providing accurate data of the field.

The IoT has attracted attention of the European Commission, yielding the project of CERP-IoT (Cluster of European Research Projects on the IoT). This project is one of the world's most active projects in the field of the Internet of Things. It focuses on applications of the Internet of Things in the field of industry and environment, and of society at large. One of the noteworthy projects of the field of IoT is IoT-A (The Internet of Things Architecture), funded by European EP7. Apart from IoT-A, the European EP7 also runs projects like IoT@Work and IoTi (Internet of Things Initiative), which both focus on development of new IoT infrastructure benefiting several areas of human lives. With the aim of letting the earth speak for itself, the Hewlett-Packard enterprise runs CeNSE (Central Nervous System for the Earth). The project aims to design an effective IoT system using billions of sensing devices and distributing them throughout the earth to achieve their goal. The secondary aim of the project is to develop smaller, cheaper, and more robust sensing devices which can detect minor vibrations and motions on the earth. This ambitious project aims to cover almost the whole earth with IoT infrastructure.

14.7 Significant Challenges

In contrast to many benefits of the Internet of Things, the concept has a few severe challenges to overcome. With the growing actualization of the IoT, the concerns related to such challenges have also gained the attentions of research communities and governments [4]. Following are a few of these challenges:

1. **Efficient identity management of IoT devices:** Due to the wide spread of the IoT in almost all facets of human life, there is fast growth in devices connected through the Internet. Current the IP address system cannot manage billions of devices effectively, and hence development of a new naming convention and system identity management system is badly needed. Such a system should be able to allocate exclusive identities to the rapidly growing number of Internet-connected devices.

2. **Standardization and interoperability:** Currently, different manufacturers use exclusive technologies and services in their devices that may not be accessible by devices of other manufacturers. For effective and global infrastructure of the IoT, interoperability among all objects and sensors is inevitable, which can be brought on only by standardization of IoT.

3. **Security and safety of objects:** As the IoT connects a large network of objects that might be spread over a large geographical area and enables them to exchange data, there is a very big threat to the IoT infrastructure to get targeted by impostors who may intend to get hold of that supposedly personal/confidential information or may corrupt the data being communicated or may cause damage to the objects physically. Thus, security and safety of the objects is of paramount importance.

4. **Privacy, encryption, and secure network capable of being programmed:** The majority of the IoT uses simple Radio Frequency Identification (RFID) and two-dimensional barcodes for identifying the objects involved in the IoT infrastructure. As these technologies are very common and hence vulnerable in terms of data privacy, measures have to be taken in order to ensure privacy and secured programming-enabled networks. Due to the interdependency of the data producing/collecting section and the data processing section of the IoT infrastructure, it is essential that sensors are equipped with adequate encryption systems to ensure the integrity of the data at the data processing section of IoT infrastructure.

5. **Optimization of energy usage by IoT:** As a large number of devices may share and process data at their end using electricity and the Internet, a large amount of energy will be used by the IoT infrastructure.

With the growing spread of the IoT concept, it will be inevitable to develop a cutting-edge technology to minimize the usage of the energy by the IoT. The option of green technology also needs to be well explored to yield energy-efficient IoT devices.

6. **Dedicated spectrums:** As the IoT infrastructure may connect billions of devices to communicate a wide range of data, a dedicated spectrum for safe and secured data transfer is essential. By growth of the IoT infrastructure at global and local level, this requirement will need primary attention.

14.8 Future of Internet of Things Network

In this section, the future of the Internet of Things network along with its details is presented. A vision of the future IoT network is described in Figure 14.1. The details regarding each component will be elaborated in this section.

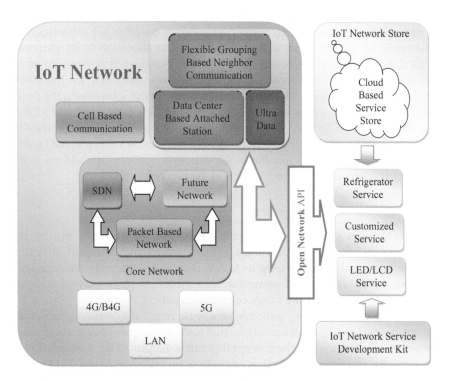

FIGURE 14.1
Outline of upcoming IoT network.

14.8.1 Software-Defined Network (SDN)

A software-defined network (SDN) empowers the users for programming the switch. SDN mainly empowers programming of network service devices. OpenFlow is a popular software-defined network. It enables users to design a switch through coding that can alter the protocol and can also check new protocols. For coding the switch to alter the protocol and for testing a new protocol, one may use this programmable network (PN).

14.8.2 IoT Device Management

For the devices connected to IoT infrastructure, ideally there should be support from the IoT network by a mechanism of plug-and-play for the IoT devices. Currently, devices are manually being connected to the network by their users. However, for effective IoT infrastructure, it is desirable that devices get automatically configured without user intervention. For this, a plug-and-play mechanism is essential for the IoT network.

Another important issue is efficient allocation of IP addresses, through which devices can connect to the IoT infrastructure and effectively communicate data. It seems that the present IP address assignment scheme doesn't work if the IoT involves a large number of devices. Looking to the growth of the concept at a global level and in every facet of human life, it is very much possible that almost all devices of daily use may get connected to one or the other IoT infrastructure. This could be a challenging task with reference to allocation of IP addresses to all these devices. For efficient handling of IoT-connected devices, an effective and mountable addressing scheme for connecting devices to the Internet is needed. An IPv6 address allocation scheme is one of the solutions to the problem. There are two scopes for unicast address, namely links local and global. The first one, local link, is used for auto-discovery and auto-configuration. It doesn't assure a unique address in the larger networks. Routers will not be able to forward it to other links too. For larger networks, global scope address is the solution. It ensures address uniqueness in the large networks.

14.8.3 Connection Management

Communication protocol possessed by different devices may not be the same. Hence, the connection management object may support different standards to nodes that belong to users. Different standards are managed by Access Point (AP) or IoT home gateway. However, there may be an issue when the node is not in the range of communication.

In case of mobile phones, which have a wide range of communication, they cannot be installed in small sensors, as that leads to a higher price and excessive battery usage. Grouping is the concept that can be helpful in the management of connectivity of the devices. Such devices are of two types: physical and logical.

Figure 14.2 explains both types of groupings in the IoT infrastructure. Physical grouping connects devices that are physically close to each

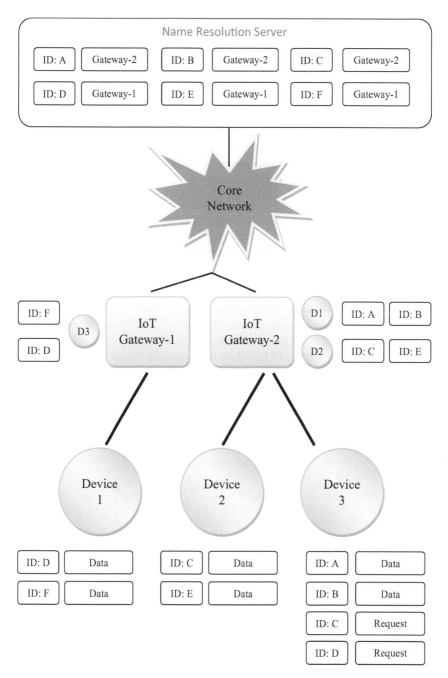

FIGURE 14.2
The concept of physical grouping and logical grouping.

other. Connection among equal IoT service devices is required to support service-oriented networks without generating direct connections with each other.

14.9 IoT Gateway

The IoT gateway is the core component in the future IoT network. Figure 14.3 presents the overview of the IoT gateway.

The following subsection describes the key features to compose the IoT gateway.

14.9.1 Network-to-Network (N2N) Communication Support

In the composition of IoT gateways, the important part is the management of information provided by concerned nodes' analysis of contents and network properties available in the cluster of objects that utilize content specific network (CSN) or distributed information network (DSN). Moreover, these kinds of information need to be accessed by information management servers or peer network objects to facilitate information-based multi-connection or information-based routing strategy. A network object communicates with

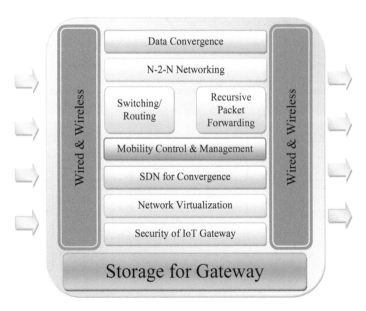

FIGURE 14.3
A glance at an IoT gateway.

other network objects through the core network for sharing information of the network. But the flip side is that sharing of information through newly designed connections of network objects may lead to problems of overhead, which might result in performance degradation in the core network. In this scenario, a technique is required to transmit the data without reconfiguring the new resulting networks among the network to solve this problem. Resultantly, protocol is required for network-to-network communication in an IoT gateway, which can act as an object for network management.

14.9.2 Routing Techniques

Routing algorithms play an important role not only in the routing for subnetworks of IoT infrastructure but also the routing of the related core network. A sensor kind of network that acts as a multi-hop network can be assigned the task of routing for the internal network that is the group of nodes or objects in the same subnet. This technique appears to be sufficient for primary IoT nodes comprised of sensors. However, there is still a problem of implementation in the IoT environment. First, there is an assumption in IoT architecture that there are chances of high mobility of sensors or nodes. Based on this assumption, the device pertaining to the user may be connected to the nearest object of the sub-network when the user is in a mobile state. Specifically, certain devices like sensors have a greater chance than other devices to connect to the nearest sub-network because of limited battery resources. Based on the limitation, a network group may break when a device tries to connect to a nearby network object, and then managing the network becomes chaotic. Thus, a routing algorithm is required for all the devices or objects to present them from connecting to different physical groups.

A possible remedy to tackle this challenge can be an algorithm that considers identification (ID) of things instead of an IP-based routing algorithm from the core network. A scenario depicting ID-based algorithm is shown in Figure 14.4.

Assume that a device labeled as 1 requires contents whose identifiers are C and D. Device 1 posts a request to the IoT gateway No. 2 that the contents of C and D are needed. After this, Gateway No. 2 recognizes that device 2 maintains content labeled as D. There are three methods by which Gateway No. 2 can obtain the content of D.

The first solution is that network-to-network communication with IoT gateway No. 1 is performed by Gateway No 2. So, it posts a request to Gateway No 1 for the content D. However, this solution will not work when two gateways are situated at too-far locations. Another solution is to request the name resolution server that maintains numerous identifications and locations of content. In this case, Gateway No 2 can fetch the information that Gateway No 1 has: the information about the location of content D. This method is easy to implement but overhead related to Gateway No 2 posts

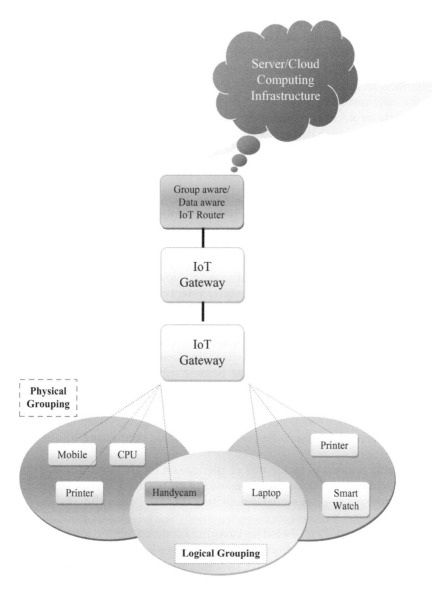

FIGURE 14.4
A scenario depicting an ID-based algorithm.

the request message to neighboring networks. On receiving the request message by Gateway No. 1, it replies along with content to Gateway No. 2 by applying backward technique. This illustration is simple for directly embedding in the core network, so a lot of new studies are requesting implementation of content-based routing techniques.

14.9.3 Mobility Support

Mobility is an important area of concern in contemporary networking and communication techniques. The various devices of communication have the property of mobility, and different studies have guided mobility support in IP networks and cellular networks. In the case of cellular networks, the issue of mobility is solved by a mobility management server that maintains the location of the network. However, this is feasible only with a large number of servers and infrastructure. Moreover, devices of cellular networks consume battery resources for connectivity.

In IP networking, mobility support is proposed by solutions like Mobile IP. But implementation of these solutions is challenging due to message authentication issues, delays in connection and the complex management of IP. Thus, mobility support in an IP network is required by the client when connection is lost. Even though it is an easy step, enormous switchover and connectivity are compromised. Moreover, performance and the quality of service (QoS) of the network seems to be degraded. Therefore, a new concept of mobility support is needed. Also, unlike cellular networks, IP networking can support mobility by network management objects in terms of robustness rather than creating new infrastructure and a large number of servers.

14.9.4 Packet Flow

In the IoT environment, lots of packets for communication are generated, since targetless connections are required to address the problem of delay. It is time-consuming to identify appropriate nodes for content and posting a connection request message; for example, to control the temperature of the room, each sensor must be able to read the temperature within the room. This information is to be sensed by the sensors, and messages are forwarded to the controlling server in the existing network. The controlling server analyses the received data and makes the connection established to send the temperature control order. So it takes numerous steps just to control the temperature of the room.

In the IoT environment, it is important to handle the issues of delay for a targetless connection. Moreover, multicasting or packet-forwarding techniques are suitable for this type of connection because these techniques facilitate sending of multiple packets to multiple parallel networks. But many packets for communication are generated, and techniques to improve network throughput are needed.

14.9.5 Software Defined Network

At the end level of the network in an IoT environment, there is a requirement of software-defined network, as every device requires a different network rate to fulfill the QoS, and guest devices demanding connection to the

IoT gateway need to be organized. The data generated by the server and actuators of the current network and data from different networks may be used by content. In this scenario, an IoT gateway must have the ability for convergence of data for speedy collection of data and remove the obsolete data or content. The processing of huge amounts of data and tailored data, as per user need, is the key for successful implementation of an IoT gateway. Therefore, data collection from different sources, converging it with the sensor network's information and converting it into transmittable data are important features of an IoT network.

14.9.6 Security Features in IoT Network

An IoT gateway provides an interface between a IoT network and an IoT device. All data pertaining to IoT devices is preserved by IoT gateways, and any kind of query can be solved by an IoT gateway as data of IoT devices is already with the IoT gateway. In this situation, confidentiality of data is not an issue for IoT devices. However, confidentiality of data is to be maintained during IoT gateway and IoT device communication. An IoT gateway has an algorithm; to ensure privacy, a novel security algorithm for privacy would go a long way toward implementing security in an IoT gateway [9].

Further wireless modes of communication between IoT gateways and IoT devices poses a problem of eavesdropping. Also, an IoT device is based on limited power—a battery—so implementation of a security algorithm may not be a viable idea. Rather, a lightweight security algorithm is required to secure the IoT data.

Registration of IoT nodes and their authenticity also needs to be address in the process of securing an IoT gateway. Depending on the type of node, like master node, member node or guest node, various kinds of registration types are required for communication with the gateway.

14.10 Conclusion

This chapter introduced the concept of the "Internet of Things," which is a futuristic approach to connect all things in a single network. The sensor device provides intelligence to the IoT network for communication and transmission of data, and it supports intelligent decisions. The various types of communication, like between humans, between humans and devices, and between device and device, are automated through the IoT network. This chapter illustrates architecture for future IoT networks and techniques for this architecture to facilitate management of the IoT network and its various nodes as well as grouping and privacy.

This chapter also focused on different feasible future applications and various country-specific projects related to the IoT and challenges in implementing IoT techniques. The deployment of the IoT could be tough and demands active research efforts, but it can lead to significant impact in terms of professional and financial benefit in the coming years.

References

1. A. M. Turing, "Computing Machinery and Intelligence," *Mind*, Volume LIX, Issue 236, 1 October 1950, pp. 433–460, doi:10.1093/mind/LIX.236.433.
2. S. Madakam, R. Ramaswamy, S. Tripathi, "Internet of Things (IoT): A literature review," *Journal of Computer and Communications*, 3, 164–173, 2015.
3. D. D. Sanju, A. Subramani, & V. K. Solanki, "Smart City: IoT Based Prototype For Parking Monitoring & Parking Management System Commanded By Mobile App," *Second International Conference on Research in Intelligent and Computing in Engineering*. Gopeshwar, Uttrakhand, India, March 24–26, 2017.
4. V. K. Solanki, M. Venkatesan, & S. Katiyar, "Conceptual model for smart cities for irrigation and highway lamps using IoT," *International Journal of Interactive Multimedia and Artificial Intelligence*, 4(3), 28–33, 2017.
5. R. Khan, S. U. Khan, R. Zaheer, & S. Khan, "Future Internet: The Internet of Things Architecture, Possible Applications and Key Challenges," *10th International Conference on Frontiers of Information Technology*, pp. 257–260. IEEE, 2012.
6. V. K. Solanki, M. Venkatesan, & S. Katiyar, "Think Home: A Smart Home as Digital Ecosystem in Circuits and Systems," *Circuits and Systems*, 7(8), 1976–1991, 2018.
7. R. Dhall & V. Solanki, "An IoT Based Predictive Connected Car Maintenance Approach," *International Journal of Interactive Multimedia and Artifiial Intelligence*, 3(4), 16–22, 2017.
8. V. Kadam, S. Tamane, & V. Solanki, "Smart and Connected Cities through Technologies," *IGI-Global*. doi:10.4018/978-1-5225-6207-8.
9. V. K. Solanki, S. Katiyar, V. B. Semwal, P. Dewan, M. Venkatesan, & N. Dey, *"Advance Automated Module for Smart and Secure City"* in ICISP-15, organized by G. H. Raisoni College of Engineering & Information Technology, Nagpur, Maharashtra, on December 11–12, 2015, published by Procedia Computer Science, Elsevier, 78, 367–374, 2016.

15

Impact of IoT to Accomplish a Vision of Digital Transformation of Cities

Shagun Tyagi, Mihir Joshi, Nabila Ansari, and V. K. Singh

CONTENTS

15.1 Introduction

Today communication through the Internet is done between humans, but when considering future internet objects that can communicate through themselves and back, that interaction becomes the Internet of Things (IoT). The communication forms will expand from human to things to machine-to-machine interaction. The Internet of Things (IoT) (Botterman, 2009; Ishida and Isbister, 2000) is a trending concept of this age, grabbing a lot of attention of researchers and academicians. Before going into an in-depth meaning of IoT, let's just start with an overview of the origin of IoT. The term "IoT" was initially used in 1999 by Kevin Ashton, who is a well-known British technology pioneer, with a view to describe a system in which objects can be connected to the Internet by sensors or tiny actuators (Mulani and Pingle, 2016).

IoT devices currently have a wide scope of applications, including connected security systems, smart healthcare devices, smart energy meters, thermostats, wearable devices, smart home, smart cars, smart electronic appliances and smart lights in household and commercial environments, alarm clocks, speaker systems, vending machines, and more, essentially to accomplish a vision of smart cities. There is no single commonly accepted definition of the IoT. Different authors have given their own perspectives on IoT. The Internet of Things is an interweaving of networks, from the user end to virtual reality, and within devices. Every device is interconnected with other devices through IP addresses assigned to these objects so that synchronization and communication could be maintained (Li et al., 2011; Reaidy et al., 2015). With help of Information and communications technology (ICT), we can enhance the quality and performance of urban services for the formation of smart cities (Solanki et al., 2016).

On the basis of the preceding discussion, it can be concluded that the term "IoT" means a pattern where countless devices employ communication services offered by Internet protocols. A large number of these devices, frequently called "smart objects," are not directly used by people, but rather exist as parts in structures or vehicles and are spread out in the environment. IoT allows people and things to connect anytime by using the Internet through any device—whether a mobile phone, laptops, computer, or any other—from any distance. For example, working couples who are not home all the time to monitor basic things like water supply can use IoT devices to manage it from anywhere, with the help of a technology called WaterSenz. Worldwide usage of the Internet, to any service from any location, is becoming more accessible (Mulani and Pingle, 2016).

The primary contributions of this chapter are as follows:

1. Review of issues and challenges in the implementation of IoT in an Indian scenario
2. Revisiting the communication network model for the device-to-device link
3. A hypothesized model for saving electricity
4. Futuristic approach to IoT for smart cities
5. Blockchain approach and challenges in India

15.2 Issues and Challenges Associated with IoT Adoption

It is extensively accredited that the IoT advancements and applications are still in their emerging stages (Miorandi et al., 2012). There still are many issues and challenges associated with IoT, and they require more research such as technology, legal and regulatory rights, normalization, reasonably priced queries and infrastructural development, data access, data security and privacy issues, information systems management, and internet connectivity (Atzori et al., 2010; Miorandi et al., 2012).

The Internet of Things establishes a connection between the virtual and real worlds, and becomes aware of context and can sense, communicate, interact, and exchange data, information, and knowledge from one place to another. The information and data can be transferred from one device to another at any time and from any distance. Several base algorithms can be created and applied to the software applications for better IoT interfaces.

1. **Information security issues:** The overall security and pliability of the Internet of Things is a function of how security risks are assessed and managed. IoT services provide new security and privacy challenges in everyday life, indicating the user's level of accepting any application. The SSL or secure socket layer security provided by third-party security seal providers will soon become obsolete with the introduction of blockchains, or asymmetric cryptography. Users will prefer applications that are risk-free (Shin, 2014). Low-risk-taker users will be willing to pay to secure their devices. Security of a device is a function of the risk that a device will be compromised, the damage such compromise will cause, and the time and resources required to achieve a certain level of protection. IoT is expanding the number of connected devices integrated into our everyday lives, presenting the opportunity for cyber attackers to gain access to our physical world through security holes in these new systems.

2. **User privacy issues:** Privacy of users' data seems to be the biggest concern with the explosion of IoT devices. Lack of trust in IoT restricts one's decision to connect with IoT devices. As IoT enables numerous daily things to be followed, observed, and associated, a great deal of individual and public data can be gathered automatically (Atzori et. al., 2010). An individual expects privacy and the fair use of his or her data. Due to deployment, mobility, and complexity in IoT, it is difficult to use (Wan and Jones, 2013). Protecting privacy in the IoT environment becomes more serious as the number of attack vectors on its information and communication technology (ICT) entities appears to be growing (Li et al., 2011; Roman et al., 2011; Ting and IP, 2015). For example, a smart car as an IoT device could keep records of passengers and their tours, distance covered by the car and its behavior, and traffic patterns. Cab companies can utilize this data to understand rush-hour time; the government can use this data to solve traffic congestions. It is a smarter way to interchange information and solve the problems of the public but compromises public privacy, which therefore requires better policies.

3. **Calibration issues:** Calibration in IoT aims to lower the entry barriers for new service providers and users, to improve the compatibility of different applications/systems, and to allow products or services to better perform at a higher level. In a fully compatible environment, any IoT device would be able to connect to any other device or system and exchange information as desired. In practicality,

interoperability is more complex. Information is exchanged at different levels in IoT devices and to different degrees (Miorandi et al., 2012). Regulation of IoT devices is important so that a fine-tuning could be established among devices for coordination, as the communication of IoT devices is not limited to one point or one country. Various standards used in IoT (e.g., security standards, communication standards, and identification standards) might be the key enablers for the spread of IoT technologies and need to be designed to embrace emerging technologies. Specific issues in IoT calibration include the ability to exchange and use information, radio access level issues, semantic ability to exchange and use information, and security and privacy issues (Wang et al., 2013).

4. **Legal and regulatory issues:** As the systems connected with the IoT become advanced, with more complex interfaces, the legality of these systems will also become more demanding. Civil society will need stronger regulations to manage the data flow and sharing of the data across systems. This requires a legal framework for handling the civil rights of the citizens, which will severely influence the associated policies related to the IoT applications and digital environments. The potential threat to the systems from hackers and AI bots needs more advanced security framework and rules.

5. **Emerging economy and development issues:** The spread and effect of the Internet are worldwide in nature, giving opportunity and advantages to create online communities and discussion forums. In the meantime, there is a pattern of difficulties arising in varying areas identified with the sending, advancing, implementing, and utilization of innovation. It is sensible to anticipate that those difficulties will also be true regarding the potential advantages and difficulties related to the Internet of Things. Coincidentally, McKinsey Global Institute noticed that IoT innovation has a huge potential to create economies.

6. **Technical issues and challenges:** Designing SOA for IoT is a big challenge, in which benefit-based things may experience the ill effects of execution and cost constraints. Adaptability is perilous at various levels, including information exchange and systems administration, information preparing and administration, and administration provisioning (Miorandi et al., 2012). It is a challenging task, to create organizing innovations and norms that can permit information assembled by an extensive number of gadgets to move proficiently inside IoT systems. Likewise, intense administration disclosure techniques and object-naming services should be created to spread IoT innovation (Atzori et al., 2010).

15.3 IoT Communication Methods and Applications

A device-to-device (D2D) communication model can be designed as a system that is with or without human intervention. The basic method involves establishing a connection within the devices through short-range communication networks vis-à-vis WiFi, Bluetooth, and RFID; and through long-range communication networks vis-à-vis Wi-MAX and mobile communication. The end-to-end communication among the devices can be single hop or multiple hops. In single-hop communication, the devices communicate through a base station, which is called the network infrastructure. However, in multiple hops, communication is established end to end between the devices. The D2D communication can be set up under the licensed spectrum and unlicensed spectrum (Bello and Zeadally, 2016). To establish a successful and unified IoT communication process, the presence of smart things is important, such as RFID tags, sensors, and so on (Chaves and Decker, 2010) that can not only store the data but also transmit and reflect sensitivity in doing so.

These sensors and low-cost actuators installed into the devices are the initial parameters to create an IoT. It also depends on the transmission power, connectivity software (that requires timely up gradation and debugging), availability of memory, and heterogeneity of devices. A successful IoT implementation requires a constant flow of information between the heterogeneous devices across various networks. Relaying data is important for real-time and anywhere accessibility. If the base station or the focal transmitter fails to establish a connection with the device, as it usually occurs during the loss of signal, D2D communication can relay the data and hence maintain the connectivity of the devices. This is the basic difference between normal connectivity establishment by the devices and the IOT communication procedure, whereby the device itself formulates the necessary arrangement for securing the connection. Mobile mechanization, clearing control, and accessing connectivity points is the basis of intelligence measurement in D2D communication. Establishing synchronization of the devices with other devices that may not be similar in specifications is another task that is expected of the devices. Further, to transmit and retransmit the data received, the devices may not require base-station connection.

Application of this concept in smart cities could be to water pumps used domestically, whereby the sensor will communicate with the mobile device when the pump has remained on for a very long duration, so as to save electricity. However, if the sensor notes that no communication could be established on a mobile device, then it may cut the electricity supply to the pump. This whole procedure could be recorded over the cloud and the data may be secured by email. Thus, the user will be able to know how much electricity is being consumed by each device (Figure 15.1).

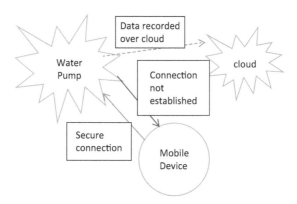

FIGURE 15.1
IoT and cloud-enabled electricity-saving mechanism.

Additionally, the IoT communication process in a device-to-device or device-to-cloud occurs over various layers. The OSI model, or the Open Systems Interconnection model, is the basis of communication that consists of two basic layers: the upper layer, also known as higher layers, and lower layers. The upper layers are nearer to the user and handle the encryption and decryption of the data. This layer establishes the inter-host communication and manages the data flow within the applications. The lower layers lay out the data received from the higher layers in the form of digital or analog signals. The two base layers have seven different sub-layers of communication. Each sub-layer has its own Protocol Data Unit (PDU) over which the data is transmitted. Each layer performs an encapsulation process of the data so as to transfer from one layer to the other without the loss of data segments. A short description of these layers is as follows:

1. The physical layer (PHY) transmits the signals in analog or digital form. The communicating devices receive and transmit the data in the form of streaming bits. This layer is responsible for the transmission of signals over optical fiber cables or radio electric networks. It may be executed through a PHY chip which is embedded in Ethernet device, a Wi-Fi device or 3G/4G/LTE, and Wi-Max devices.

2. The data link layer is responsible for the maintaining the communication between the two devices that are directly linked with each other. These devices may also be connected to another device that helps in establishing a connection. For that purpose, the MAC addresses are assigned to the devices for communication. This layer transmits and receives the data in the form of frames. The data link layer works over short-range or long-range communication networks like Zigbee, Wi-Fi, Bluetooth, and Ethernet, all of which are widely used in IoT.

3. The network layer performs the logical addressing of the machines, managing connections between machines and routing of the data in the form packets. This layer regulates the path taken by the data and forms a step-by-step communication. The network layer works on the IPV4 or IPV6 protocol.

4. The transport layer maintains the flow control of the data in the form of segments or datagram. The connection is established through a port and the device for an end-to-end data transmission. The most generally applied protocol on this layer is TCP, or the transmission control protocol.

5. The session layer manages the sessions between the applications over which the data segments have been transmitted. It establishes and maintains the inter-host communication that coordinates with the applications. This layer follows half-duplex or full-duplex operation. Application environs that utilize Remote Procedure Calls (RPCs) are incorporated in the session layer. The commonly applied protocols are Netbios, ASP, and more.

6. The presentation layer handles the encryption and decryption of the data transmitted or received. It transfigures the machine data into a format that can be operated by other systems. A number of different protocols are used in application layer: ASCII, Unicode, SMB, AFP, and others.

7. The application layer provides an access to the network for the transmission of the data. It is also known as the abstraction layer, and it divides codes into different sections. The most widely used protocol in the application layer for IoT is HTTP; other protocols such as Gopher, SMTP, FTP, and Telnet are also used extensively in the implementation of the application layer.

The TCP/IP or Transmission Control Protocol/Internet Protocol model constitutes of five layers in which the application, presentation, and session layers have been summated as one single layer called the application layer. The physical layer in TCP/IP follows Wi-Fi, Wi-Max, Lora, and NFC protocols. It describes the nature of communication cabled (fiber optics) or radio frequency communication and the encoding or coding procedure and wavelengths for signal transmission.

The *data link layer* handles the packet transmissions over the physical layer with specifically encoded bits for the start and end of the data frames. The protocols followed on the data link layer are Ethernet, Token Ring, and Wireless Ethernet. The data frames consist of encoded bits that points to a specified machine over which the data is transmitted.

The basic function of the network layer is to transmit the data packets and route them across multiple networks. This layer is therefore also named the internet layer, since IPV4 or IPV6 (Internet Protocol Version 4 and 6) are used extensively to establish a network connection between the machines by identifying the host computer and pointing the location. The IPV4 is the 32-bit version, whereas the IPV6 is the 128-bit version. Each and every protocol has a unique number for interpretation, like Internet Control Message Protocol (ICMP) and Internet Group Management Protocol (IGMP), which are assigned numbers 1 and 2. Both these protocols lie above the network layer but coordinate with the network layer for routing and host address identification.

In order to ensure that the data has been delivered to the appropriate destination, the *transport layer* constitutes of all those dynamic protocols that maintain the correct ordering of the data packets. For example, the SYN bit is used to make the connection, whereas the ACK bit is used to acknowledge the connection establishment. Initially, SYN=1 and ACK=0 that show the connection request made, which shows that there is no prior request in the piggyback. The connection establishment shows SYN=1 and ACK=1, which shows that SYN displays both the connection request and connection accepted, but ACK is used to differentiate between the two.

The application layer encapsulates the data according to the codes used for securing the data packets to be transmitted across the transport layer. This layer is a combination of the session, presentation and application layers of the OSI model. It utilizes protocols for host configuration and routing such as Dynamic Host Configuration Protocol (DHCP), Hyper Text Transfer Protocol (HTTP), File Transfer Protocol (FTP), and others.

15.4 Blockchain Approach and Challenges

In order to resolve the Byzantine General problem, Satoshi Nakamoto in 2008 created the blockchain methodology and crypto-currency. Well-known examples of crypto-currency are Bitcoin and Ethereum. The ideology behind blockchain creation is to create a trustworthy environment for financial transactions, reducing costs, improvisations to securing databases, and making devices and systems impermeable to external threats. Figures 15.2 and 15.3 display the hypothesized model for application of blockchains in electronic voting machines (EVMs) through Aadhaar Card or UID cards. Figure 15.3 shows the implementation of blockchain for securing transactions, processing information in linkages with the financial agencies (bank accounts or crypto-currency). The usage of blockchains is to limit or terminate the presence of third-party security seal providers and third-party database management. The concept of blockchain is a symmetrical process that involves the use of hashing methods and a cryptographic algorithm. The purpose of blockchains is to decentralize the database and

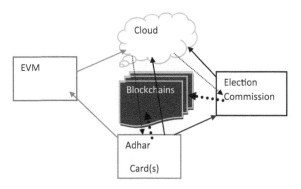

FIGURE 15.2
UID or Aadhaar Card and EVM blockchain.

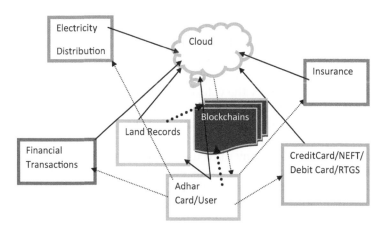

FIGURE 15.3
Blockchain for Aadhaar and citizen necessities.

convert it into a distributed frame. The initial block or the genesis block determines the beginning of the blockchains.

As a representation, the Bitcoin blockchain works on four core steps involving Wallet, Storing, Mining, and Routing. The Wallet provides the necessary cryptic keys that provide access for transactions to the user. Storing keeps a replica of the block in the node so that it can be further accessed and propagated. The Mining function generates new blocks in the chain using the miner nodes, that is, the nodes that perform work proofing. Proof of work is an essential computational task performed by the miners. The network verifies the new block created by the miner. The propagation of blocks and transactions occur over the network between the users is known as Routing. Block creation is a trustless consensus mechanism and highly expensive process for any attacker to generate a malicious block, as the nodes generating cryptic blocks will outrun the attacker in time and motion (Sheetal and Venkatesh, 2018). Such blockchain method is called a permissionless blockchain, used in Bitcoin and Ethereum, where the blocks can be viewed, transactions can be changed, and nodes can be created and left without permission; however, these are protected by cryptic algorithms.

In the hypothesized models in Figures 15.2 and 15.3, the two major issues have been represented:

1. EVMs or electronic voting machines, a technology that is facing questions of tampering and changes in the votes input by citizens; and

2. Aadhaar Card or Unique Identification Card, linking with all services necessary for the users.

The blockchain technology includes two distinct methods: permissioned and permissionless platforms. The EVM can be connected with the Aadhaar Card portal that gives unique access to the voter. The cryptic algorithm shall identify the user biometric details and verify the location of the voter. Once a genesis block is created by the vote, then it should be accessible only to the voter and the Election Commission office. The user can cast the vote through a UID online or through a mobile platform. The vote drafted by the user shall be encrypted and stored till the next vote comes in. The permissioned mechanism will not allow any third user to view, modify, or hack the blockchain that would be connected to the cloud for storage and data analysis. In Figure 15.3 the UID or the Aadhaar Card is connected to all the basic necessities of a citizen. The electricity usage or billing can be linked to the user's Aadhaar card so as to avoid the loss of transmission due to electricity theft; remote areas can be connected easily without any third-party contracting. In the same way, insurance is an important aspect, as the linkages with the Aadhaar Card widen the scope for the insurance firms to identify the user will be efficient and validation would be easy to notify the assets (cars, homes, etc.). Further, the Aadhaar card link with the blockchain can enable authorities to assess the land buy or sale records, land lease, commercial or not-for-profit activities, and even allow authorities to track the smuggling activities in a forest area. In smart city projects, the implementation of blockchains will revolutionize the payment interfaces and reduce security breach problems. Likewise, credit card or debit card companies will be able to provide more secure services to customers by eliminating the need for third-party security, thus reducing costs of transactions by making Aadhaar card as a source of the medium of transaction (Reyna et al., 2018). However, the implementation of this technology poses legal challenges besides the availability of infrastructure. Moreover, acquisition of information from users is a question, since most of the users may find it unreasonable to share information through IoT-enabled devices. Policy creation is another hurdle in the Indian context, where the consensus needs to be built within the executive. Imparting education to make people aware of IoT and IOeT systems is another challenge that authorities should undertake to reduce the gap between technology and user.

15.5 Role of IoT in Digital Transformation of Cities

With the expanding number of the total populace moving to and living in urban areas, there has been a keen interest in the development of smart communities. There is an emergence of smart cities because energy and resources are saved in the smart city infrastructure (Kadam et al., 2019). Smart cities begin with a smart public infrastructure to convey clean water, reliable power, safe gas, and productive open lighting. Also, they free up assets by keenly conveying fundamental administrations, putting resources into different sections to enhance personal satisfaction. The quick development of brilliant urban communities has reignited enthusiasm on the theme and produced much discourse. As the examination on smart cities extends, new thoughts and terms, for example, smart countries, are added to the literature. A smart city can be defined as a city that has the ability to control and supervise conditions of all of its critical infrastructure—water, power, roads, airports, rails, subways, communications, and buildings—in order to track and optimize its resources and to create preventive maintenance plans while maximizing services to its citizens. Additionally, problems like traffic congestions, pollution, or shortage of energy supplies can be solved by using IoT applications.

The smart city mission is to improve the old urban infrastructure in cities. The concept of smart cities is to make their existence in the development of urban development policies. There are lot of existing technologies in urban cities, such as the Internet, wireless communication, infrared, Bluetooth, and Wi-Fi, varying among technologies and their range. A smart city makes optimal use of public resources by increasing the quality of services (Kumar et al., 2017). The impact of the smart city on society is obtained by creating sustainable economic development and increasing the quality of life by exceeding in other multiple areas like mobility, economy, living, and government: for example, smart belly trash, smart parking, smart lighting, smart monitoring of devices, smart roads and transportation, smart healthcare, and more (Kim et al., 2017). IoT connects not only humans but also connects a machine to humans and back. The innovation advances should make the city, the planet, and the society a healthy, happy place for people to live, learn, and grow. Due to implementation of IoT, living standards and quality of life for people is improving; however, this empirical research on the smart city and its connection with IoT is still comparatively low (Kim et al., 2017; Lee, et al., 2014).

Initially, the idea is to lead the development of a smart city and encourage innovation to create competitive advantage. Smart cities are also viewed as the sustainable development factor since the late 1990s, when global warming issues added to business priorities. Cyber connection of cities augmented the terminologies such as intelligent cities or super-intelligence units (Komninos, 2002).

A large number of features are covered in smart cities that are linked directly or indirectly to the life of people. These areas comprise of not only man-made resources (infrastructure such as transportation) but access to natural resources. Besides these aspects, smart cities enhance societal competitiveness by improving the living standards, society participation, and human capital. The IoT and smart cities concept also involve governance, matters of economy, as well as tools such as 3D printing, robotics, and real-time big data analysis (Carvalho, 2015).

Sensors and wireless networks help to link physical objects to the Internet, creating a digital world. This Internet of Things, whereby everyday devices are connected and share data through a network, will have a significant effect on society. Traditionally, computing power resided in mainframe computers, and then it moved to a client-server architecture, where it is distributed between two parts. With cloud-based computing, on-demand computing capability is now easily available through the Internet. The Internet of Things shall soon become widespread, not only to connect to in the future, but also to interconnect terminals like mobile phones, computers, smart devices, and everyday life objects. Besides, it will also help in setting up communication from human to human (H2H) and turning into machine to machine (M2M) communications (Chen et al., 2012). IoT research and development is gaining momentum in the industry and academic community recently.

IoT can also contribute to securing rail routes, resolve fencing issues at international boundaries, secure safety at sensitive zones like oil rigs and mines, and manage maintenance of equipment. IoT in a smart city plays a crucial part. Sensor nodes developed for associating IoT-enabled devices allow information to move from one point to another. This information is shared with multiple departments of the city, and authorities can manage, correct, and provide information suitable for fulfilling their purpose.

The interconnected department will give a better arrangement, and it diminishes the possibility of occurrence of any issue by controlling and tackling it. IoT can be put into action at public places, streets and activities, groups, business complexes, workplaces, and different utilities departments. IoT can help in providing benefits to the general public by deducing basic problem solutions. It can act as a source of knowledge to create frameworks and help keep a self-check on the environment. Cloud computing helps to store the dynamic data coming continuously from different machine sources that are connected with IoT devices. IoT tracks all the constant exercises occurring around it, processing and storing them over the cloud. This huge information framework helps to simplify treatment of the information. IoT can be an independent framework or coordinate with other units to work in parallel. This automated process is then given support by the actuator, holding up the contribution from the sensor hub of the IoT.

IoT is such a complex network to put into action in a city that it converts into the Internet of Everything (IOeT). Executing such a complex network doesn't become lacuna free. One of the real issues subsequent to actualizing such a structure is security and support. The system of IoT must be for all intents and purposes impervious by any unapproved individual. Although there is no way to get unauthorized access into the IoT devices, hackers may find ways to intrude into the network. IoT connections move from node to node. One node giving incorrect information can prompt a jumble of conclusive data and create serious issues. Furthermore, along these lines, the IoT arrangement must be secured in all these perspectives. The information being exchanged must be scrambled and institutionalized, keeping in mind the end goal: to keep up its respectability. The information coming and running must be marked with legitimate keys, keeping in mind the end goal is to keep up its vividness. Standard cryptography conventions and strategy will beat the issue of security and design required to actualize an IoT. A smart city uses and reuses its waste, utilizing it over and over. More mechanical headway and development in science prompts new techniques for development of the current method for living. In any case, to accomplish these objectives, city specialists need to embrace activities and systems that make the physical-computerized condition of smart cities, actualizing helpful applications and e-benefits and guaranteeing the long-haul supportability of keen urban communities through practical business models. To provide citizen access to a wide array of services like home appliances, weather monitoring, temperature assessment, pollution measurement, surveillance, vehicle security, home automation, medical equipment, traffic systems, and others, device management is essential. System administration has to manage inflow and outflow of the enormous amount of data from individuals, organizations, and executives (Bellavista et al., 2013; Sundmaeker et al., 2010). The IoT may manifest itself within a number of products and services spread across a diverse range of fields, from intelligent management of factories and firms to self-driven machines.

15.6 Smart Solutions for Smart Cities

The literature on smart cities displays several innovative ideas that have segmented the nature of the advance of digital properties. The locus of IoT is on advance computing and computational methods, complex algorithm designs, wireless networks, and public areas (Komninos, 2002 and 2008). Furthermore, IoT increases the transparency of government regulations, omnipresence of data, and entrée of different types of information, improvising individual thoughts associated to the city, like maintaining hygiene and cleanliness of surroundings, cooperating with public administration, and empowering citizen involvement in city matters (Cuff et al., 2008).

The vision of IoT is applicable not only through advance technological interfaces but also through machine–human understanding. The researcher discusses a few of the applications of IoT technology such as IoT Energy Assessment Device: The advance IoT setups give rise to the use of renewable energy resources, such as an extensive implementation of solar panels or photovoltaic cells (Zanella et al., 2014). The implementation of IoT devices will effectively manage the energy usage of the whole city, street light management, control cameras, pollution check, or air quality evaluation. This, in turn, will help to control the electricity production that can be applied over the power framework of the city. Globally, the objective of leadership is to reduce the dependency on fossil fuels, the introduction of advance systems utilizing bio-fuels, hydropower, biomass, geothermal power generation, and to use oceans as a power source. Such procedures will give rise to sustainable behavior within the society. Additionally, the IoT applications can be best applied to transportation and logistics. The actuators and sensors are fixed on intelligent mobile systems that can predict the best suitable path for a vehicle, a way to avoid congestion of transport, parking facilities, service lane management, timely arrival and departure of local transport buses or metros, and so on. IoT can be utilized appropriately in managing the transportation systems of the cities and enhance the security features that can be applied to vehicles. A fully automated home is also an advance example of IoT application, wherein the house maintains cleanliness, security, temperature control of interiors, and intensity of light, as per the user's convenience. The objective is to bring enough relaxation to the user by understanding the user mood and necessity. The devices may be connected to LAN or Ethernet connectivity or Wi-Fi enabled.

The sensors and transducers connected to these devices will provide real-time access to the data given by users, like facial changes and emotion recognition, and will monitor the security of the interior and exterior of the house. Further, the smart homes may attain access to the network in case of loss of transmission through the Neighbour Area Network (NAN). Through this, the smart homes can share data and send or receive signals regarding matters of extreme safety. Various organizations are working in the direction to introduce one device for multiple functions (Lin et al., 2017). Advances in medicine and digitizing medical records generate a lot of help to the IoT systems, which can thus be utilized to provide emergency services to the residents. The IoT system establishes communication with the smart homes so as to keep patients, infants, and the elderly under close observation (Shafiekhah et al., 2016). Similarly, the ageing of national heritage sites and monuments can be tracked. The theft of historical artifacts and exemplary works of art and crafts from the past can be protected using the IoT. The database maintained over the cloud by IoT devices can provide timely assistance and maintenance of historical works.

Additionally, the database will also assist in predicting natural calamities and disasters like earthquakes and volcanic activity, so that the public can remain well aware and be protected in advance. IoT devices can calculate the real-time health data of an individual by assessing his or her heart rate, pulse rate, sugar levels, cardiovascular disease, or stroke-like situations. Furthermore, IoT devices can well be utilized in waste management, which is a prime challenge in the urban area. The collection of garbage from sources and delivering to the destination costs more, and the service delivery uses a lot of time. The application of IoT to waste management can help introduce cost-saving schemes for loading and recycling of the waste. An optimal solution to the management of waste-carrying trucks and routing to the available locations for dumping can help manage the city waste (Nuortio et al., 2006). IoT-enabled and -connected devices can provide better solutions to water management by addressing the wastage of drinking water, alerting authorities about the chemical levels, assessing the pollution level in water, assessing the Chemical Oxygen Demand (COD) and Biological Oxygen Demand (BOD) in the water bodies, water pressure management in the pipelines, and keeping track of flood-like situations. IoT devices can also be used to keep in check the noise pollution created by the devices. The sensors or the actuators can help reduce the noise dissipation by the devices so that if the device crosses a certain decibel level, it may then itself be able to reduce the noise emission (Zanella et al., 2014). Currently, there are no specific apps that have been made to monitor all the devices at home, due to obvious inability of the devices to connect with the Internet. However, once all the devices are connected with the Internet, managing devices at home or from the office or any other place will become easier and efficient. The data will be saved on the cloud, and the devices can be accessed through the cloud. Users can get access to the data from anywhere and even synchronize it with the other service providers, which gives the user full power to manipulate and utilize data.

Another area where the IoT adoption is necessary and important for assessing is the environmental fluctuations. The changes in weather and climatic conditions have changed the activities in nature. Lately, tsunamis and other similar natural disasters across the globe have made early warning systems a necessity. IoT can be easily connected to mobile devices and cover large geographical locations. By assessing the parameters such as wind speed, humidity, barometric pressure, and other changes in weather, the devices can act as active handheld warning systems. These devices can also check the rise in pollution levels and simultaneously, the decrease in pollution levels after the planting of trees. Therefore, IoT devices can check soil, air, and water quality.

Although several measures have been employed by city administrations globally, traffic management remains a great hurdle for the civil life. IoT can provide real-time solutions to traffic situations by projecting appropriate pathways to the commuter. Apart from handheld devices, even a microwave oven, if linked to Wi-Fi, can predict traffic situations while in the kitchen. It can also make several suggestions to the user for which road to take and how to avoid heavy traffic based on the data collected over time. The traffic signals can be connected to these devices that can show the real-time videos to the citizen for choosing the perfect road for mobility and can help the traffic administration to put traffic in order. The smart lighting can adjust the intensity of the light in the daytime and in the night based on the changing weather conditions. These sensors can also switch the source of power from different sources like solar panels or electricity so as to keep power consumption in check (Zanellaet et al., 2014). A software called Park Smart offers the citizen perfect parking areas, based on Radio Frequency Identification (RFID) and Near Field Communication (NFC). These sensors give citizens a legitimate entry so that they can park in their own slots, and also the sensor may direct the user to an available space. The real-time access to the data through video signals thus directs the driver to the nearest parking location.

15.7 Issues and Challenges with Smart Cities

Most technologists and designers are occupied with examining how to build smart cities and what features to give them. But on the other side, it's important to know who is going to live in these cities and what needs to be required to be a citizen of the smart city. There are also lots of difficulty in the development of smart cities such as effective waste management, scarcity of resources in a country such as sufficient electricity, traffic congestions, and inadequate ageing infrastructures that need to be modified, maintaining air quality, and other problems (Borja, 2007). ICT is one of the most important factors required for building smart cities. They can improve the administration and working of a city. Regardless of declared favorable circumstances and advantages of ICTs' use in urban areas, their effect is as yet indistinct. Indeed, they improve quality of life but also promote inequality (Odendaal, 2003). The fundamental difficulties that urban communities look for in the smart governance activity field are connected with the pressing requirement for a change in government models. Governance policies must be more flexible. Two other main challenges are demographic changes and territorial cohesion. There so many issues related to the natural environment too, as there is a reduction in energy consumption, increasing pollution, and CO_2 emissions from vehicles. These environmental issues also influencing the growing ecological demand for achieving a sustainable development. Poverty and urban insecurity are the other major challenges in this field. Financing is likewise one of the real difficulties in the development of smart cities, as are activity clogs, ICT foundation shortfall, cyber security, and shortages in access to innovation. The fundamental destinations of smart city ventures must be to take care of urban issues in an effective method to enhance the manageability of the city and quality of life of its inhabitants.

15.8 Futuristic Approach of IoT for Creating Smart Cities

With the discovery and development of the IoT and cheap sensors has come an increase in the potential of smart cities, and now a vision of smart cities is becoming an original reality. Combining cities with smart tech solutions like remote control home appliances, smart automated vehicles, sensor-based devices, and so on has become the expectation of today's digital age. IoT service-providing corporations are CISCO, IBM, Apple, SIEMENS, and others. Countries like India, the Netherlands, France, South Korea, and the United States are working on building advance systems useful for establishing smart cities. These nations are adopting IoT solutions with the aim to deliver citizens greater safety and security, more effective water management, more efficient energy usage, smarter parking and traffic management, a convenient and stress-free lifestyle, and an overall standard living for those living in urban areas, to make available to them the latest technology. This will be more clear with a few examples: Let's suppose you are going to the cinema or maybe for to a shopping center, and your car drops you off at your desired destination and then drives itself to a nearby parking space and notifies you via your phone that it is safely parked in a parking area. A large number of shopping centers or public gathering places, parks, and car parking lots need to be established for bringing changes in the cities for IoT advancements.

Another example could be considered are web applications that allow citizens to report any emergency, like an accident, directly to a nearby hospital or ambulance for help. Sensors monitoring road and weather conditions could provide real-time updates on any potential environmental change for an unexpected rainfall, thunderstorm, and so on, or detect people throwing garbage or smoking in restricted areas. Many such items could bring great benefits to a smart city environment.

Smart cities and IoT innovation solutions will impact nations, but still, the adoption rate among the nations is relatively low due to lack of trust and not having sufficient technology. To make sure this new technology works for all, governments should make effective policies. For this governments must first support the development of 5G networks, by giving required incentives and liberty to the telecom industry. Investing in networks of the future will provide the necessary infrastructure to release technologies such as 5G and life-improving IoT services. With these life-enhancing services in place, smart cities and IoT will lead the charge to improve quality of life and living standards and help in creating and evolving the new sectors for employment. Overall, what is clear is that smart cities powered by IoT are inward bound, and those who are in charge of managing and running them need to start developing plans. By looking ahead and thinking about the future today, the policies, infrastructure, and technologies that need to be designed will be developed and implemented in the least disruptive and most cost-effective way possible.

15.9 Conclusion and Implications

With advancements in technology, changes in society are inevitable. The next generation will enter an ultra-modern stage that will bring about everything in a mechanized form. IoT implementation in ultra-modern cities will give rise to AI and other allied impervious machines that will help society to advance into the next decade. It is interesting to know that when a billion devices get connected to one another, their interaction will change the way the current generation handles machines. In this chapter, the researchers have reviewed the current IoT technology and also presented the next generation of IoT systems. The primary focus of IoT systems according to the literature is on enhancing citizens' lifestyles by providing better health facilities, advance safety measures in the domestic vicinity, understanding human behavior through facial recognitions, employing more AI for public protection, and making cities sustainable for human survival. This chapter started with the definitions and introduction to the IoT systems and also covered the issues and vulnerabilities of IoT systems. The issues discussed in this chapter largely focused on security and privacy of the data, calibration of the data, and legal technological issues. The more advanced the systems get, the more complicated will become the implementation of these systems.

Further, this chapter also discussed the models required for the inter-operations and communications of the IoT systems. These communication network models are OSI and TCP/IP that actually mobilize the IoT systems by establishing the uplinks with the Internet. Various protocols were also discussed in brief to revisit and understand the protocol nature of the networking structure. Additionally, this chapter proposed a framework for optimal use of electricity by connecting all devices to the IoT that can deduce the amount of electricity consumed in a household. The role of IoT in digital transformation of the cities has also been discussed, accentuating the necessity of connecting devices with the Internet so that the maintenance of the city infrastructure, roads, rails, and communication channels can help resolve everyday issues such as traffic congestion, vehicular pollution monitoring, smart hospitals, smart trauma centers, and tracking illegal or criminal activities. Moreover, segmentation of this chapter contains practical applications of IoT smart solutions for making smart cities.

References

Atzori, L., Iera, A., and Morabito, G., 2010. The internet of things: A survey. *Computer Networks*, 54(15), pp. 2787–2805.

Bellavista, P., Cardone, G., Corradi, A., and Foschini, L., 2013. Convergence of MANET and WSN in IoT urban scenarios. *IEEE Sensors Journal*, 13(10), pp. 3558–3567.

Bello, O., and Zeadally, S., 2016. Intelligent device-to-device communication in the internet of things. *IEEE Systems Journal*, 10(3), pp. 1172–1182.

Borja, J. 2007. Counterpoint: Intelligent cities and innovative cities. *Universitat Oberta de Catalunya (UOC) Papers: E-Journal on the Knowledge Society*, 5. Available from http://www.uoc.edu/uocpapers/5/dt/eng/mitchell.pdf.

Botterman, M. 2009. Internet of Things: An early reality of the future Internet. *Report of the Internet of Things Workshop*. European Commission Information Society and Media Directorate General, Prague.

Carvalho, L. 2015. Smart cities from scratch? A socio-technical perspective. *Cambridge Journal of Regions, Economy and Society*, 8, 43–60.

Chaves, L. W. F., and Decker, C., 2010, June. A survey on organic smart labels for the Internet-of-Things. In *Networked Sensing Systems (INSS), 2010 Seventh International Conference* on (pp. 161–164). IEEE.

Chen, M., Wan, J., and Li, F., 2012. Machine-to-machine communications: Architectures, standards and applications. *KSII Transactions on Internet & Information Systems*, 6(2).

Cuff, D., Hansen, M., and Kang, J., 2008. Urban sensing: Out of the woods. *Communications of the ACM*, 51(3), pp. 24–33.

Ishida, T., and Isbister, K. (Eds.). 2000. *Digital Cities: Technologies, Experiences, and Future Perspectives*. Berlin, Germany: Springer Science & Business Media.

Kadam, V. G., Tamane, S. C., and Solanki, V. K., 2019. Smart and connected cities through technologies. In *Big Data Analytics for Smart and Connected Cities*. (pp. 1–24). IGI Global.

Kim, T. H., Ramos, C., and Mohammed, S. 2017. Smart city and IoT. 159–162.

Komninos, N. 2002. *Intelligent Cities: Innovation Knowledge Systems and Digital Spaces*. London, UK: Routledge.

Komninos, N., 2008. *Intelligent Cities and Globalisation of Innovation Networks*. London, UK: Routledge, pp. 137–172.

Kumar, A. D. V., Malarchelvi, P. D. S. K., and Arockiam, L., 2017. CALDUEL: Cost and load overhead reduction for route discovery in load protocol. *Advances in Computer and Computational Sciences*. Springer, pp. 229–237.

Lee, J. H., Hancock, M. G., and Hu, M. C., 2014. Towards an effective framework for building smart cities: Lessons from Seoul and San Francisco. *Technological Forecasting and Social Change*, 89, 80–99.

Li, X., Lu, R., Liang, X., Shen, X., Chen, J., and Lin, X. 2011. Smart community: An internet of things application. *IEEE Communications Magazine*, 49(11).

Lin, T., Rivano, H., Le Mouël, F., Baron, B., Spathis, P., Rivano, H., de Amorim, M. D. et al., 2017. A survey of smart parking solutions. *Parking*, 1524, 9050.

Miorandi, D., Sicari, S., De Pellegrini, F., and Chlamtac, I., 2012. Internet of things: Vision, applications and research challenges. *Ad hoc Networks*, 10(7), pp. 1497–1516.

Mulani, T. T., and Pingle, S. V., 2016. Internet of things. *International Research Journal of Multidisciplinary Studies*, 2(3).

Nuortio, T., Kytöjoki, J., Niska, H., and Bräysy, O., 2006. Improved route planning and scheduling of waste collection and transport. *Expert Systems with Applications*, 30(2), pp. 223–232.

Odendaal, N. 2003. Information and communication technology and local governance: Understanding the difference between cities in developed and emerging economies. *Computers, Environment and Urban Systems*, 27(6), 585–607.

Reaidy, P. J., Gunasekaran, A., and Spalanzani, A., 2015. Bottom-up approach based on Internet of Things for order fulfillment in a collaborative warehousing environment. *International Journal of Production Economics*, 159, 29–40.

Reyna, A., Martín, C., Chen, J., Soler, E., and Díaz, M. 2018. On blockchain and its integration with IoT. Challenges and opportunities. Future Generation Computer Systems, 88, 173–190.

Roman, R., Najera, P., and Lopez, J., 2011. Securing the internet of things. *Computer*, 44(9), pp. 51–58.

Shafie-Khah, M., Kheradmand, M., Javadi, S., Azenha, M., de Aguiar, J. L. B., Castro-Gomes, J., Siano, P., and Catalão, J. P. S., 2016. Optimal behavior of responsive residential demand considering hybrid phase change materials. *Applied Energy*, 163, pp. 81–92.

Sheetal, M., and Venkatesh, K. A., 2018. Necessary requirements for blockchain technology and its applications. *International Journal of Computing Science and Information Technology*, Special Issue, pp. 130–133.

Shin, D., 2014. A socio-technical framework for Internet-of-Things design: A human-centered design for the Internet of Things. *Telematics and Informatics*, 31(4), pp. 519–531.

Solanki, V. K., Katiyar, S., BhashkarSemwal, V., Dewan, P., Venkatasen, M., and Dey, N. 2016. Advanced automated module for smart and secure city. *Procedia Computer Science*, 78, 367–374.

Sundmaeker, H., Guillemin, P., Friess, P., and Woelfflé, S., 2010. Vision and challenges for realising the internet of things. *Cluster of European Research Projects on the Internet of Things, European Commission*, 3(3), pp. 34–36.

Ting, S. L., and IP, W. H., 2015. Combating the counterfeits with web portal technology. *Enterprise Information Systems*, 9(7), pp. 661–680.

Wan, J., and Jones, J. D., 2013. Managing IT service management implementation complexity: From the perspective of the Warfield version of systems science. *Enterprise Information Systems*, 7(4), pp. 490–522.

Wang, F., Ge, B., Zhang, L., Chen, Y., Xin, Y., and Li, X., 2013. A system framework of security management in enterprise systems. *Systems Research and Behavioral Science*, 30(3), pp. 287–299.

Zanella, A., Bui, N., Castellani, A., Vangelista, L., and Zorzi, M. Internet of things for smart cities. *IEEE Internet of Things journal*, 2014, 1(1), 22–32.

Index

Note: Page numbers in italic and bold refer to figures and tables, respectively.